THE FAT OF
THE LAND

MORE WILDSIDE CLASSICS

THE FAT OF THE LAND

The Story of an American Farm

JOHN WILLIAMS STREETER

WILDSIDE PRESS

THE FAT OF THE LAND

This edition published 2005 by Wildside Press, LLC.
www.wildsidepress.com

CHAPTER I

My Excuse

My sixtieth birthday is a thing of yesterday, and I have, therefore, more than half descended the western slope. I have no quarrel with life or with time, for both have been polite to me; and I wish to give an account of the past seven years to prove the politeness of life, and to show how time has made amends to me for the forced resignation of my professional ambitions. For twenty-five years, up to 1895, I practised medicine and surgery in a large city. I loved my profession beyond the love of most men, and it loved me; at least, it gave me all that a reasonable man could desire in the way of honors and emoluments. The thought that I should ever drop out of this attractive, satisfying life, never seriously occurred to me, though I was conscious of a strong and persistent force that urged me toward the soil. By choice and by training I was a physician, and I gloried in my work; but by instinct I was, am, and always shall be, a farmer. All my life I have had visions of farms with flocks and herds, but I did not expect to realize my visions until I came on earth a second time.

I would never have given up my profession voluntarily; but when it gave me up, I had to accept the dismissal, surrender my ambitions, and fall back upon my primary instinct for diversion and happiness. The dismissal came without warning, like the fall of a tree when no wind shakes the forest, but it was imperative and peremptory. The doctors (and they were among the best in the land) said, "No more of this kind of work for years," and I had to accept their verdict, though I knew that "for years" meant forever.

My disappointment lasted longer than the acute attack; but, thanks to the cheerful spirit of my wife, by early summer of that year I was able to face the situation with courage that grew as strength increased. Fortunately we were well to do, and the loss of professional income was not a serious matter. We were not rich as wealth is counted nowadays; but we were more than comfortable for ourselves and our children, though I should never earn another dollar. This is not the common state of the physician, who gives more and gets less than most other men; it was simply a happy combination of circumstances. Polly was a small heiress when we married; I had some money from my maternal grandfather; our income was larger than our necessities, and our investments had been fortunate. Fate had set no wolf to howl at our door.

In June we decided to take to the woods, or rather to the country, to see what it had in store for us. The more we thought of it, the better I liked the plan, and Polly was no less happy over it. We talked of it morning, noon, and night, and my half-smothered instinct grew by what it fed on. Countless schemes at length resolved themselves into a factory farm, which should be a source of pleasure as well as of income. It was of all sizes, shapes, industries, and limits of expenditure, as the hours passed and enthusiasm waxed or waned. I finally compromised on from two hundred to three hundred acres of land, with a total expenditure of not more than $60,000 for the building of my factory. It was to produce butter, eggs, pork, and apples, all of best quality, and they were to be sold at best prices. I discoursed at some length on farms and farmers to Polly, who slept through most of the harangue. She afterward said that she enjoyed it, but I never knew whether she referred to my lecture or to her nap.

If farming be the art of elimination, I want it not. If the farmer and the farmer's family must, by the nature of the occupation, be deprived of reasonable leisure and luxury, if the conveniences and amenities must be shorn close, if comfort must be denied and life be reduced to the elemental necessities of food and shelter, I want it not. But I do not believe that this is the case. The wealth of the world comes from the land, which produces all the direct and immediate essentials for the preservation of life and the protection of the race. When people cease to look to the land for support, they lose their independence and fall under the tyranny of circumstances beyond their control. They are no longer producers, but consumers; and their prosperity is contingent upon the prosperity and good will of other people who are more or less alien. Only when a considerable percentage of a nation is living close to the land can the highest type of independence and prosperity be enjoyed. This law applies to the mass and also to the individual. The farmer, who produces all the necessities and many of the luxuries, and whose products are in constant demand and never out of vogue, should be independent in mode of life and prosperous in his fortunes. If this is not the condition of the average farmer (and I am sorry to say it is not), the fault is to be found, not in the land, but in the man who tills it.

Ninety-five per cent of those who engage in commercial and professional occupations fail of large success; more than fifty per cent fail utterly, and are doomed to miserable, dependent lives in the service of the more fortunate. That farmers do not fail nearly so often is due to the bounty of the land, the beneficence of Nature, and the ever-recurring seed-time and harvest, which even the most thoughtless cannot interrupt.

The waking dream of my life had been to own and to work land; to own it free of debt, and to work it with the same intelligence that has made me successful in my profession. Brains always seemed to me as necessary to success in farming as in law, or in medicine, or in business. I always felt that mind should control events in agriculture as in commercial life; that listlessness, carelessness, lack of thrift and energy, and waste, were the factors most potent in keeping the farmer poor and unreasonably harassed by the obligations of life. The men who cultivate the soil create incalculable wealth; by rights they should be the nation's healthiest, happiest, most comfortable, and most independent citizens. Their lives should be long, free from care and distress, and no more strenuous than is wholesome. That this condition is not general is due to the fact that the average farmer puts muscle before mind and brawn before brains, and follows, with unthinking persistence, the crude and careless traditions of his forefathers.

Conditions on the farm are gradually changing for the better. The agricultural colleges, the experiment stations, the lecture courses which are given all over the country, and the general diffusion of agricultural and horticultural knowledge, are introducing among farming communities a more intelligent and more liberal treatment of land. But these changes are so slow, and there is so much to be done before even a small percentage of our six millions of farmers begin to realize their opportunities, that even the weakest effort in this direction may be of use. This is my only excuse for going minutely into the details of my experiment in the cultivation of land. The plain and circumstantial narrative of how Four Oaks grew, in seven years, from a poor, ill-paying, sadly neglected farm, into a beautiful home and a profitable investment, must simply stand for what it is worth. It may give useful hints, to be followed on a smaller or a larger scale, or it may arouse criticisms which will work for good, both to the critic and to the author. I do not claim experience, excepting the most limited; I do not claim originality, except that most of this work was new to me; I do not claim hardships or difficulties, for I had none; but I do claim that I made good, that I arrived, that my experiment was physically and financially a success, and, as such, I am proud of it, and wish to give it to the world.

I was fifty-three years old when I began this experiment, and I was obliged to do quickly whatever I intended to do. I could devote any part of $60,000 to the experiment without inconvenience. My desire was to test the capacity of ordinary farm land, when properly treated, to support an average family in luxury, paying good wages to more than the usual number of people, keeping open house for many friends, and at the same time not depleting my bank account. I wished to experiment in *intensive farming*, using ordinary farm land as other men might do under similar or modified circumstances. I believed that if I fed the land, it would feed me. My plan was to sell nothing from the farm except finished products, such as butter, fruit, eggs, chickens, and hogs. I believed that best results would be attained by keeping only the best stock, and, after feeding it liberally, selling it in the most favorable market. To live on the fat of the land was what I proposed to do; and I ask your indulgence while I dip into the details of this seven years' experiment.

You may say that few persons have the time, inclination, taste, or money to carry out such an experiment; that the average farmer must make each year pay, and that the exploiting of this matter is therefore of interest to a very limited number. Admitting much of this, I still claim that there is a lesson to every struggling farmer in this narrative. It should teach the value of brain work on the farm, and the importance of intelligent cultivation; also the advantages of good seed, good tilth, good specimens of well-bred stock, good food, and good care. Feed the land liberally, and it will return you much. Permit no waste in space, product, time, tools, or strength. Do in a small way, if need be, what I have done on a large scale, and you will quickly commence to get good dividends. I have spent much more money than was really necessary on the place, and in the ornamentation of Four Oaks. This, however, was part of the experiment. I asked the land not only to supply immediate necessities, but to minister to my every want, to gratify the eye, and please the senses by a harmonious fusion of utility and beauty. I wanted a fine country home and a profitable investment within the same ring fence.

Will you follow me through the search for the land, the purchase, and the tremendous house-cleaning of the first year? After that we will take up the years as they come, finding something of special interest attaching naturally to each. I shall have to deal much with figures and statistics, in a small way, and my pages may look like a school book, but I cannot avoid this, for in these figures and statistics lies the practical lesson. Theory alone is of no value. Practical application of the theory is the test. I am not imaginative. I could not write a romance if I tried. My strength lies in special detail, and I am willing to spend a lot of time in working out a problem. I do not claim to have spent this time and money without making serious mistakes; I have made many, and I am willing to admit them, as you will see in the following pages. I do claim, however, that, in spite of mistakes, I have solved the problem, and have proved that an intelligent farmer can live in luxury on the fat of the land.

CHAPTER II

The Hunting of the Land

The location of the farm for this experiment was of the utmost importance. The land must be within reasonable distance of the city and near a railroad, consequently within easy touch of the market; and if possible it must be near a thriving village, to insure good train service. As to size, I was somewhat uncertain; my minimum limit was 150 acres and 400 the maximum. The land must be fertile, or capable of being made so.

I advertised for a farm of from two hundred to four hundred acres, within thirty-five miles of town, and convenient to a good line of transportation. Fifty-seven replies came, of which forty-six were impossible, eleven worth a second reading, and five worth investigating. My third trip carried me thirty miles southwest of the city, to a village almost wholly made up of wealthy people who did business in town, and who had their permanent or their summer homes in this village. There were probably twenty-seven or twenty-eight hundred people in the village, most of whom owned estates of from one to thirty acres, varying in value from $10,000 to $100,000. These seemed ideal surroundings. The farm was a trifle more than two miles from the station, and 320 acres in extent. It lay to the west of a north-and-south road, abutting on this road for half a mile, while on the south it was bordered for a mile by a gravelled road, and the west line was an ordinary country road. The lay of the land in general was a gentle slope to the west and south from a rather high knoll, the highest point of which was in the north half of the southeast forty. The land stretched away to the west, gradually sloping to its lowest point, which was about two-thirds of the distance to the western boundary. A straggling brook at its lowest point was more or less rampant in springtime, though during July and August it contained but little water.

Westward from the brook the land sloped gradually upward, terminating in a forest of forty to fifty acres. This forest was in good condition. The trees were mostly varieties of oak and hickory, with a scattering of wild cherry, a few maples, both hard and soft, and some lindens. It was much overgrown with underbrush, weeds, and wild flowers. The land was generally good, especially the lower parts of it. The soil of the higher ground was thin, but it lay on top of a friable clay which is fertile when properly worked and enriched.

The farm belonged to an unsettled estate, and was much run down, as little had been done to improve its fertility, and much to deplete it. There were two sets of buildings, including a house of goodly proportions, a cottage of no particular value, and some dilapidated barns. The property could be bought at a bargain. It had been held at $100 an acre; but as the estate was in process of settlement, and there was an urgent desire to force a sale, I finally secured it for $71 per acre. The two renters on the farm still had six months of occupancy before their leases expired. They were willing to resign their leases if I would pay a reasonable sum for the standing crops and their stock and equipments.

The crops comprised about forty acres of corn, fifty acres of oats, and five acres of potatoes. The stock was composed of two herds of cows (seven in one and nine in the other), eleven spring calves, about forty hogs, and the usual assortment of domestic fowls. The equipment of the farm in machinery and tools was meagre to the last degree. I offered the renters $700 and $600, respectively, for their leasehold and other property. This was more than their value, but I wanted to take possession at once.

CHAPTER III

The First Visit to the Farm

It was the 8th of July, 1895, when I contracted for the farm; possession was to be given August 1st. On July 9th, Polly and I boarded an early train for Exeter, intending to make a day of it in every sense. We wished to go over the property thoroughly, and to decide on a general outline of treatment. Polly was as enthusiastic over the experiment as I, and she is energetic, quick to see, and prompt to perform. She was to have the planning of the home grounds—the house and the gardens; and not only the planning, but also the full control.

A ride of forty-five minutes brought us to Exeter. The service of this railroad, by the way, is of the best; there is hardly a half-hour in the day when one cannot make the trip either way, and the fare is moderate: $8.75 for twenty-five rides,—thirty-five cents a ride. We hired an open carriage and started for the farm. The first half-mile was over a well-kept macadam road through that part of the village which lies west of the railway. The homes bordering this street are of fine proportions, and beautifully kept. They are the country places of well-to-do people who love to get away from the noise and dirt of the city. Some of them have ten or fifteen acres of ground, but this land is for breathing space and beauty—not for serious cultivation. Beyond these homes we followed a well-gravelled road leading directly west. This road is bordered by small farms, most of them given over to dairying interests.

Presently I called Polly's attention to the fact that the few apple trees we saw were healthy and well grown, though quite independent of the farmer's or the pruner's care. This thrifty condition of unkept apple orchards delighted me. I intended to make apple-growing a prominent feature in my experiment, and I reasoned that if these trees did fairly well without cultivation or care, others would do excellently well with both.

As we approached the second section line and climbed a rather steep hill, we got the first glimpse of our possession. At the bottom of the western slope of this hill we could see the crossing of the north-

and-south road, which we knew to be the east boundary of our land; while, stretching straight away before us until lost in the distant wood, lay the well-kept road which for a good mile was our southern boundary. Descending the hill, we stopped at the crossing of the roads to take in the outline of the farm from this southeast corner. The north-and-south road ran level for 150 yards, gradually rose for the next 250, and then continued nearly level for a mile or more. We saw what Jane Austen calls "a happy fall of land," with a southern exposure, which included about two-thirds of the southeast forty, and high land beyond for the balance of this forty and the forty lying north of it. There was an irregular fringe of forest trees on this southern slope, especially well defined along the eastern border. I saw that Polly was pleased with the view.

"We must enter the home lot from this level at the foot of the hill," said she, "wind gracefully through the timber, and come out near those four large trees on the very highest ground. That will be effective and easily managed, and will give me a chance at landscape gardening, which I am just aching to try."

"All right," said I, "you shall have a free hand. Let's drive around the boundaries of our land and behold its magnitude before we make other plans."

We drove westward, my eyes intent upon the fields, the fences, the crops, and everything that pertained to the place. I had waited so many years for the sense of ownership of land that I could hardly realize that this was not another dream from which I would soon be awakened by something real. I noticed that the land was fairly smooth except where it was broken by half-rotted stumps or out-cropping boulders, that the corn looked well and the oats fair, but the pasture lands were too well seeded to dock, milkweed, and wild mustard to be attractive, and the fences were cheap and much broken.

The woodland near the western limit proved to be practically a virgin forest, in which oak trees predominated. The undergrowth was dense, except near the road; it was chiefly hazel, white thorn, dogwood, young cherry, and second growth hickory and oak. We turned the corner and followed the woods for half a mile to where a barbed wire fence separated our forest from the woodland adjoining it. Coming back to the starting-point we turned north and slowly climbed the hill to the east of our home lot, silently developing plans. We drove the full half-mile of our eastern boundary before turning back.

I looked with special interest at the orchard, which was on the northeast forty. I had seen it on my first visit, but had given it little attention, noting merely that the trees were well grown. I now counted the rows, and found that there were twelve; the trees in each row had originally been twenty, and as these trees were about thirty-five feet apart, it was easy to estimate that six acres had been given to this orchard. The vicissitudes of seventeen years had not been without effect, and there were irregular gaps in the rows,—here a sick tree, there a dead one. A careless estimate placed these casualties at fifty-five or sixty, which I later found was nearly correct. This left 180 trees in fair health; and in spite of the tight sod which covered their roots and a lamentable lack of pruning, they were well covered with young fruit. They had been headed high in the old-fashioned way, which made them look more like forest trees than a modern orchard. They had done well without a husbandman; what could not others do with one?

The group of farm buildings on the north forty consisted of a one-story cottage containing six rooms—sitting room, dining room, kitchen, and a bedroom opening off each—with a lean-to shed in the rear, and some woe-begone barns, sheds, and out-buildings that gave the impression of not caring how they looked. The second group was better. It was south of the orchard on the home forty, and quite near the road.

Why does the universal farm-house hang its gable over the public road, without tree or shrub to cover its boldness? It would look much better, and give greater comfort to its inmates, if it were more remote. A lawn leading up to a house, even though not beautiful or well kept, adds dignity and character to a place out of all proportion to its waste or expense. I know of nothing that would add so much to the beautification of the country-side as a building line prohibiting houses and barns within a hundred yards of a public road. A staring, glaring farm-house, flanked by a red barn and a pigsty, all crowding the public road as hard as the path-master will permit, is incongruous and unsightly. With all outdoors to choose from, why ape the crowded city streets? With much to apologize for in barn and pigsty, why place them in the seat of honor? Moreover, many things which take place on the farm gain enchantment from distance. It is best to leave some scope for the imagination of the passer-by. These and other things will change as farmers' lives grow more gracious, and more attention is given to beautifying country houses.

The house, whose gables looked up and down the street, was two stories in height, twenty-five feet by forty in the main, with a one-story ell running back. Without doubt there was a parlor, sitting room, and four chambers in the main, with dining room and kitchen in the ell.

"That will do for the head man's house, if we put it in the right place and fix it up," said Polly.

"My young lady, I propose to be the 'head man' on this farm, and I wish it spelled with a capital H, but I do not expect to live in that house. It will do first-rate for the farmer and his men, when you have placed it where you want it, but I intend to live in the big house with you."

"We'll not disagree about that, Mr. Headman."

The barns were fairly good, but badly placed. They were not worth the expense of moving, so I decided to let them stand as they were until we could build better ones, and then tear them down.

We drove in through a clump of trees behind the farm-house, and pushed on about three hundred yards to the crest of the knoll. Here we got out of the carriage and looked about, with keen interest, in every direction. The views were wide toward three points of the compass. North and northwest we could see pleasant lands for at least two miles; directly west, our eyes could not reach beyond our own forest; to the south and southwest, fruitful valleys stretched away to a range of wooded hills four miles distant; but on the east our view was limited by the fringe of woods which lay between us and the north-and-south road.

"This is the exact spot for the house," said Polly. "It must face to the south, with a broad piazza, and the chief entrance must be on the east. The kitchens and fussy things will be out of sight on the north-west corner; two stories, a high attic with rooms, and covered all over with yellow-brown shingles." She had it all settled in a minute.

"What will the paper on your bedroom wall be like?" I asked.

"I know perfectly well, but I shan't tell you."

Seating myself on an out-cropping boulder, I began to study the geography of the farm. In imagination I stripped it of stock, crops, buildings, and fences, and saw it as bald as the palm of my hand. I recited the table of long measure: Sixteen and a half feet, one rod, perch, or pole; forty rods, one furlong; eight furlongs, one mile. Eight times 40 is 320; there are 320 rods in a mile, but how much is 16-1/2. times 320? "Polly, how much is 16-1/2 times 320?"

"Don't bother me now; I'm busy."

(Just as if she could have told in her moment of greatest leisure!) I resorted to paper and pencil, and learned that there are 5280 feet in each and every mile. My land was, therefore, 5280 feet long and 2640 feet wide. I must split it in some way, by a road or a lane, to make all parts accessible. If I divided it by two lanes of twenty feet each, I could have on either side of these lanes lots 650 feet deep, and these would be quite manageable. I found that if these lots were 660 feet long, they would contain ten acres minus the ten feet used for the lane. This seemed a real discovery, as it simplified my calculations and relieved me of much mental effort.

"Polly, I am going to make a map of the place,—lay it out just as I want it."

"You may leave the home forty out of your map; I will look after that," said the lady.

In my pocket I found three envelopes somewhat the worse for wear. This is how one of them looked when my map was finished.

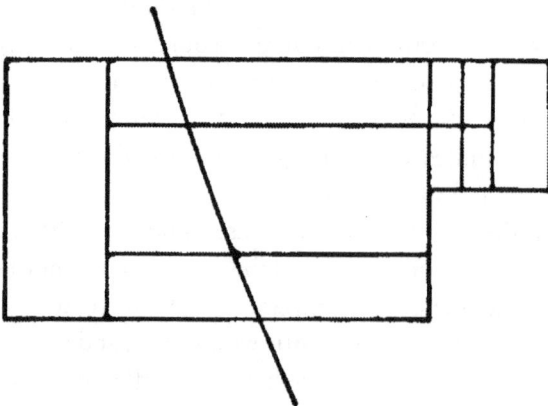

I am not especially haughty about this map, but it settled a matter which had been chaotic in my mind. My plan was to make the farm a soiling one; to confine the stock within as limited a space as was consistent with good health, and to feed cultivated forage and crops. In drawing my map, the forty which Polly had segregated left the northeast forty standing alone, and I had to cast about for some good way of treating it. "Make it your feeding ground," said my good genius, and thus the wrath of Polly was made to glorify my plans.

This feeding lot of forty acres is all high land, naturally drained. It was near the obvious building line, and it seemed suitable in every way. I drew a line from north to south, cutting it in the middle. The east twenty I devoted to cows and their belongings; the west twenty was divided by right lines into lots of five acres each, the southwest one for the hens and the other three for hogs.

Looking around for Polly to show her my work, I found she had disappeared; but soon I saw her white gown among the trees. Joining her, I said,—

"I have mapped seven forties; have you finished one?"

"I have not," she said. "Mine is of more importance than all of yours; I will give you a sketch this evening. This bit of woods is better than I thought. How much of it do you suppose there is?"

"About seven acres, I reckon, by hook and by crook; enough to amuse you and furnish a lot of wild-flower seed to be floated over the rest of the farm."

"You may plant what seeds you like on the rest of the farm, but I must have wild flowers. Do you know how long it is since I have had them? Not since I was a girl!"

"That is not very long, Polly. You don't look much more than a girl to-day. You shall have asters and goldenrod and black-eyed Susans to your heart's content if you will always be as young."

"I believe Time will turn backward for both of us out here, Mr. Headman. But I'm as hungry as a wolf. Do you think we can get a glass of milk of the 'farm lady'?"

We tried, succeeded, and then started for home. Neither of us had much to say on the return trip, for our minds were full of unsolved problems. That evening Polly showed me this plat of the home forty.

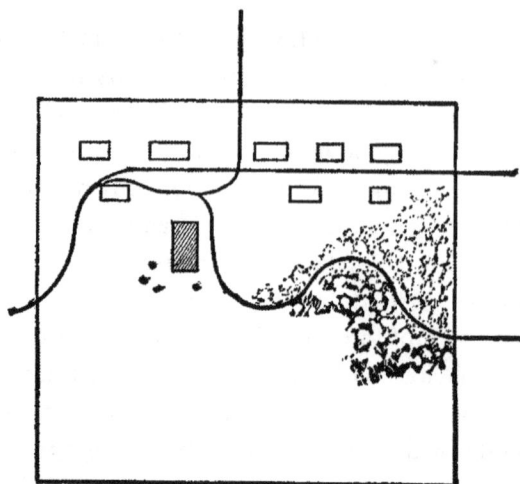

CHAPTER IV

The Hired Man

Modern farming is greatly handicapped by the difficulty of getting good help. I need not go into the causes which have operated to bring about this condition; it exists, and it has to be met. I cannot hope to solve the problem for others, but I can tell how I solved it for myself. I determined that the men who worked for me should find in me a considerate friend who would look after their interests in a reasonable and neighborly fashion. They should be well housed and well fed, and should have clean beds, clean table linen and an attractively set table, papers, magazines, and books, and a comfortable room in which to read them. There should be reasonable work hours and hours for recreation, and abundant bathing facilities; and everything at Four Oaks should proclaim the dignity of labor.

From the men I expected cleanliness, sobriety, uniform kindness to all animals, cheerful obedience, industry, and a disposition to save their wages. These demands seemed to me reasonable, and I made up my mind to adhere to them if I had to try a hundred men.

The best way to get good farm hands who would be happy and contented, I thought, was to go to the city and find men who had shot their bolts and failed of the mark; men who had come up from the farm hoping for easier or more ambitious lives, but who had failed to find what they sought and had experienced the unrest of a hand-to-mouth struggle for a living in a large city; men who were pining for the country, perhaps without knowing it, and who saw no way to get back to it. I advertised my wants in a morning paper, and asked my son, who was on vacation, to interview the applicants. From noon until six o'clock my ante-room was invaded by a motley procession—delicate boys of fifteen who wanted to go to the country, old men who thought they could do farm work, clerks and janitors out of employment, typical tramps and hoboes who diffused very naughty smells, and a few—a very few—who seemed to know what they could do and what they really wanted.

Jack took the names of five promising men, and asked them to come again the next day. In the morning I interviewed them, dismissed three, and accepted two on the condition that their references proved satisfactory. As these men are still at Four Oaks, after seven years of steady employment, and as I hope they will stay twenty years longer, I feel that the reader should know them. Much of the smooth sailing at the farm is due to their personal interest, steadiness of purpose, and cheerful optimism.

William Thompson, forty-six years of age, tall, lean, wiry, had been a farmer all his life. His wife had died three years before, and a year later, he had lost his farm through an imperfect title. Understanding machinery and being a fair carpenter, he then came to the city, with $200 in his pocket, joined the Carpenter's Union, and tried to make a living at that trade. Between dull business, lock-outs, tie-ups, and strikes, he was reduced to fifty cents, and owed three dollars for room rent. He was in dead earnest when he threw his union card on my table and said:—

"I would rather work for fifty cents a day on a farm than take my chances for six times as much in the union."

This was the sort of man I wanted: one who had tried other things and was glad of a chance to return to the land. Thompson said that after he had spent one lonesome year in the city, he had married a sensible woman of forty, who was now out at service on account of his hard luck. He also told of a husky son of two-and-twenty who was at work on a farm within fifty miles of the city. I liked the man from the first, for he seemed direct and earnest. I told him to eat up the fifty cents he had in his pocket and to see me at noon of the following day. Meantime I looked up one of his references; and when he came, I engaged him, with the understanding that his time should begin at once.

The wage agreed upon was $20 a month for the first half-year. If he proved satisfactory, he was to receive $21 a month for the next six months, and there was to be a raise of $1 a month for each half-year that he remained with me until his monthly wage should amount to $40,—each to give or take a month's notice to quit. This seemed fair to both. I would not pay more than $20 a month to an untried man, but a good man is worth more. As I wanted permanent, steady help, I proposed to offer a fair bonus to secure it. Other things being equal, the man who has "gotten the hang" of a farm can do better work and get better results than a stranger.

The transient farm-hand is a delusion and a snare. He has no interest except his wages, and he is a breeder of discontent. If the hundreds of thousands of able-bodied men who are working for scant wages in cities, or inanely tramping the country, could see the dignity of the labor which is directly productive, what a change would come over the face of the country! There are nearly six million farms in this nation, and four millions of them would be greatly benefited by the addition of another man to the working force. There is a comfortable living and a minimum of $180 a year for each of four million men, if they will only seek it and honestly earn it. Seven hundred millions in wages, and double or treble that in product and added values, is a consideration not unworthy the attention of social scientists. To favor an exodus to the land is, I believe, the highest type of benevolence, and the surest and safest solution of the labor problem.

Besides engaging Thompson, I tentatively bespoke the services of his wife and son. Mrs. Thompson was to come for $15 a month and a half-dollar raise for each six months, the son on the same terms as the father.

The other man whom I engaged that day was William Johnson, a tall, blond Swede about twenty-six years old. Johnson had learned gardening in the old country, and had followed it two years in the new. He was then employed in a market gardener's greenhouse; but he wanted to change from under glass to out of doors, and to have charge of a lawn, shrubs, flowers, and a kitchen garden. He spoke brokenly, but intelligently, had an honest eye, and looked to me like a real "find." Polly, who was to be his immediate boss, was pleased with him, and we took him with the understanding that he was to make himself generally useful until the time came for his special line of work. We now had two men engaged (with a possible third) and one woman, and my *venire* was exhausted.

Two days later I again advertised, and out of a number of applicants secured one man. Sam Jones was a sturdy-looking fellow of middle age, with a suspiciously red nose. He had been bred on a farm, had learned the carpenter's trade, and was especially good at taking care of chickens. His ambition was to own and run a chicken plant. I hired him on the same terms as the others, but with misgivings on account of the florid nose. This was on the 19th or 20th of July, and there were still ten days before I could enter into possession. The men were told to report for duty the last day of the month.

CHAPTER V

Boring for Water

The water supply was the next problem. I determined to have an abundant and convenient supply of running water in the house, the barns, and the feeding grounds, and also on the lawn and gardens. I would have no carrying or hauling of water, and no lack of it. There were four wells on the place, two of them near the houses and two stock wells in the lower grounds. Near the well at the large house was a windmill that pumped water into a small tank, from which it was piped to the barn-yard and the lower story of the house. The supply was inadequate and not at all to my liking.

My plan involved not only finding, raising, and distributing water, but also the care of waste water and sewage. Inquiring among those who had deep wells in the village, I found that good water was usually reached at from 180 to 210 feet. As my well-site was high, I expected to have to bore deep. I contracted with a well man of good repute for a six-inch well of 250 feet (or less), piped and finished to the surface, for $2 a foot; any greater depth to be subject to further agreement.

It took nearly three months to finish the water system, but it has proved wonderfully convenient and satisfactory. During seven years I have not spent more than $50 for changes and repairs. We struck bed-rock at 197 feet, drilled 27 feet into this rock, and found water which rose to within 50 feet of the surface and which could not be materially lowered by the constant use of a three-inch power-pump. The water was milky white for three days, in spite of much pumping; and then, and ever after, it ran clear and sweet, with a temperature of 54° F. Well and water being satisfactory, I cheerfully paid the well man $448 for the job.

Meantime I contracted for a tank twelve by twelve feet, to be raised thirty feet above the well on eight timbers, each ten inches square, well bolted and braced, for $430,—I to put in the foundation. This consisted of eight concrete piers, each five feet deep in the clay, three feet square, and capped at the level of the ground with a limestone two feet square and eight inches thick. These piers were set in octagon form around the well, with their centres seven feet from the middle of the bore, making the spread of the framework fourteen feet at the ground and ten at the platform. The foundation cost $32. A Rider eight-inch, hot-air, wood-burning, pumping engine (with a two-inch pipe leading to the tank, and a four-inch pipe from it), filled the tank quickly; and it was surprising to see how little fuel it consumed. It cost $215.

I have now to confess to a small extravagance. I contracted with a carpenter to build an ornamental tower, fifty-five feet high, twenty feet across at the base, and fifteen feet at the top, sheeted and shingled, with a series of small windows in spiral and a narrow stairway leading to a balcony that surrounded the tower on a level with the top of the tank. This tower cost $425; but it was not all extravagance, because a third of the expense would have been incurred in protecting the engine and making the tank frost-proof.

To distribute the water, I had three lines of four-inch pipe leading from the tank's out-flow pipe. One of these went 250 feet to the house, with one-inch branches for the gardens and lawn; another led east 375 feet, past the proposed sites of the cottage, the farmhouse, the dairy, and other buildings in that direction; while the third, about 400 feet long, led to the horse barn and the other projected buildings. From near the end of this west pipe a 1-1/2-inch pipe was carried due north through the centre of the five-acre lot set apart for the hennery, and into the fields beyond. This pipe was about 700 feet long. Altogether I used 1100 feet of four-inch, and about 2200 feet of smaller pipe, at a total cost of $803. All water pipes were placed 4-1/2 feet in the ground to be out of the reach of frost, and to this day they have received no further attention.

The trenches for the pipes were opened by a party of five Italians whom a railroad friend found for me. These men boarded themselves, slept in the barn, and did the work for seventy-five cents a rod, the job costing me $169.

Opening the sewer trenches cost a little more, for they were as deep as those for the water, and a little wider. Eight hundred feet of main sewer, a three-hundred-foot branch to the house, and short branches from barns, pens, and farm-houses, made in all about fourteen hundred feet, which cost $83 to open. The sewer ended in the stable yard back of the horse barn, in a ten-foot catch-basin near the manure pit. A few feet from this catch-basin was a second, and beyond this a third, all of the same size, with drain-pipes connecting them about two feet below the ground. These basins were closely covered at all times, and in winter they were protected from frost by a thick layer of coarse manure. They were placed near the site of the manure pit for convenience in cleaning, which had to be done every three months for the first one, once in six months for the second and rarely for the third; indeed, the water flowing from the third was always clear. This waste water was run through a drain-pipe diagonally across the northwest corner of the big orchard to an open ditch in the north lane. Opening this drain of forty rods cost $30. Later I carried this closed drain to the creek, at an additional expense of $67. The connecting of the water pipes and the laying of the sewer was done by a local plumber for $50; the drain-pipe and sewer-pipe cost $112; and the three catch-basins, bricked up and covered with two-inch plank, cost $63. The filling in of all these trenches was done by my own men with teams and scrapers, and should not be figured into this expense account. It must be borne in mind that while this elaborate water system was being installed, no buildings were completed and but few were even begun; the big house was not finished for more than a year. The sites of all the buildings had been decided on, and the farm-house and the cottage had been moved and remodelled, by the middle of October, at which date the water plant was completed. An abundant supply of good water is essential to the comfort of man and beast, and the money invested in securing it will pay a good interest in the long run. My water plant cost me a lot of money, $2758; but it hasn't cost me $10 a year since it was finished.

CHAPTER VI

We Take Possession

My barn was full of horses, but none of them was fit for farm work; so I engaged a veterinary surgeon to find three suitable teams. By the 25th of the month he had succeeded, and I inspected the animals and found them satisfactory, though not so smooth and smart-looking as I had pictured them. When I compared them, somewhat unfavorably, with the teams used for city trucks and delivery wagons, he retorted by saying: "I did not know that you wanted to pay $1200 a pair for your horses. These six horses will cost you $750, and they are worth it." They were a sturdy lot, young, well matched, not so large as to be unwieldy, but heavy enough for almost any work. The lightest was said to weigh 1375 pounds, and the heaviest not more than a hundred pounds more. Two of the teams were bay with a sprinkling of white feet, while the other pair was red roan, and, to my mind, the best looking.

Four of these horses are still doing service on the farm, after more than seven years. One of the bays died in the summer of '98, and one of the roans broke his stifle during the following winter and had to be shot. The bereaved relics of these two pairs have taken kindly to each other, and now walk soberly side by side in double harness. I sometimes think, however, that I see a difference. The personal relation is not just as it was in the old union,—no bickerings or disagreements, but also no jokes and no caresses. The soft nose doesn't seek its neighbor's neck, there is no resting of chin on friendly withers while half-closed eyes see visions of cool shades, running brooks, and knee-deep clover; and the urgent whinney which called one to the other and told of loneliness when separated is no longer heard. It is pathetic to think that these good creatures have been robbed of the one thing which gave color to their lives and lifted them above the dreary treadmill of duty for duty's sake. The kindly friendship of each for his yoke-fellow is not the old sympathetic companionship, which will come again only when the cooling breezes, running brooks, and knee-deep pastures of the good horse's heaven are reached.

A horse is wonderfully sensitive for an animal of his size and strength. He is timid by nature and his courage comes only from his confidence in man. His speed, strength, and endurance he will willingly give, and give it to the utmost, if the hand that guides is strong and gentle, and the voice that controls is firm, confident, and friendly. Lack of courage in the master takes from the horse his only chance of being brave; lack of steadiness makes him indirect and futile; lack of kindness frightens him into actions which are the result of terror at first, and which become vices only by mismanagement. By nature the horse is good. If he learns bad manners by associating with bad men, we ought to lay the blame where it belongs. A kind master will make a kind horse; and I have no respect for a man who has had the privilege of training a horse from colt-hood and has failed to turn out a good one. Lack of good sense, or cruelty, is at the root of these failures. One can forgive lack of sense, for men are as God made them; but there is no forgiveness for the cruel: cooling shades and running brooks will not be prominent features in their ultimate landscapes.

For harness and farm equipments, tools and machinery, I went to a reliable firm which made most and handled the rest of the things that make a well-equipped farm. It is best to do much of one's business through one house, provided, of course, that the house is dependable. You become a valued customer whom it is important to please, you receive discounts, rebates, and concessions that are worth something, and a community of interest grows up that is worth much.

My first order to this house was for three heavy wagons with four-inch tires, three sets of heavy harness, two ploughs and a subsoiler, three harrows (disk, spring tooth, and flat), a steel land-roller, two wheelbarrows, an iron scraper, fly nets and other stable equipment, shovels, spades, hay forks, posthole tools, a hand seeder, a chest of tools, stock-pails, milk-pails and pans, axes, hatchets, saws of various kinds, a maul and wedges, six kegs of nails, and three lanterns. The total amount was $488; but as I received five per cent discount, I paid only $464. The goods, except the wagons and harnesses, were to go by freight to Exeter. Polly was to buy the necessary furnishings for the men's house, the only stipulation I made being that the beds should be good enough for me to sleep in. On the 25th of July she showed me a list of the things which she had purchased. It seemed interminable; but she assured me that she had bought nothing unnecessary, and that she had been very careful in all her purchases. As I knew that Polly was in the habit of getting the worth of her money, I paid the bills without more ado. The list footed up to $495.

Most of the housekeeping things were to be delivered at the station in Exeter; the rest were to go on the wagons. On the afternoon of the 30th the wagons and harnesses were sent to the stable where the horses had been kept, and the articles to go in these wagons were loaded for an early start the following morning. The distance from the station in the city to the station at Exeter is thirty miles, but the stable is three miles from the city station, the farm two and a half miles from Exeter station, and the wagon road not so direct as the railroad. The trip to the farm, therefore, could not be much less than forty miles, and would require the best part of two days. The three men whom I had engaged reported for duty, as also did Thompson's son, whom we are to know hereafter as Zeb.

Early on the last day of the month the men and teams were off, with cooked provisions for three days. They were to break the journey twenty-five miles out, and expected to reach the farm the next afternoon. Polly and I wished to see them arrive, so we took the train at 1 P.M. August 1st, and reached Four Oaks at 2.30, taking with us Mrs. Thompson, who was to cook for the men.

Before starting I had telephoned a local carpenter to meet me, and to bring a mason if possible. I found both men on the ground, and explained to them that there would be abundant work in their lines on the place for the next year or two, that I was perfectly willing to pay a reasonable profit on each job, but that I did not propose to make them rich out of any single contract.

The first thing to do, I told them, was to move the large farm-house to the site already chosen, about two hundred yards distant, enlarge it, and put a first-class cellar under the whole. The principal change needed in the house was an additional story on the ell, which would give a chamber eighteen by twenty-six, with closets five feet deep, to be used as a sleeping room for the men. I intended to change the sitting room, which ran across the main house, into a dining and reading room twenty feet by twenty-five, and to improve the shape and convenience of the kitchen by pantry and lavatory. There must also be a well-appointed bathroom on the upper floor, and set tubs in the kitchen. My men would dig the cellar, and the mason was to put in the foundation walls (twelve inches thick and two feet above ground), the cross or division walls, and the chimneys. He was also to put down a first-class cement floor over the whole cellar and approach. The house was to be heated by a hot-water system; and I afterward let this job to a city man, who put in a satisfactory plant for $500.

We had hardly finished with the carpenter and the mason when we saw our wagons turning into the grounds. We left the contractors to their measurements, plans, and figures, while we hastened to turn the teams back, as they must go to the cottage on the north forty. The horses looked a little done up by the heat and the unaccustomed journey, but Thompson said: "They're all right,—stood it first-rate."

The cottage and out-buildings furnished scanty accommodations for men and beasts, but they were all that we could provide. I told the men to make themselves and the horses as comfortable as they could, then to milk the cows and feed the hogs, and call it a day.

While the others were unloading and getting things into shape, I called Thompson off for a talk. "Thompson," I said, "you are to have the oversight of the work here for the present, and I want you to have some idea of my general plan. This experiment at farming is to last years. We won't look for results until we are ready to force them, but we are to get ready as soon as possible. In the meantime, we will have to do things in an awkward fashion, and not always for immediate effect. We must build the factory before we can turn out the finished product. The cows, for instance, must be cared for until we can dispose of them to advantage. Half of them, I fancy, are 'robber cows,' not worth their keep (if it costs anything to feed them), and we will certainly not winter them. Keep your eye on the herd, and be able to tell me if any of them will pay. Milk them carefully, and use what milk, cream, and butter you can, but don't waste useful time carting milk to market—feed it to the hogs rather. If a farmer or a milkman will call for it, sell what you have to spare for what he will give, and have done with it quickly. You are to manage the hogs on the same principle. Fatten those which are ready for it, with anything you find on the place. We will get rid of the whole bunch as soon as possible. You see, I must first clear the ground before I can build my factory. Let the hens alone for the present; you can eat them during the winter.

"Now, about the crops. The hay in barns and stacks is all right; the wheat is ready for threshing, but it can wait until the oats are also ready; the corn is weedy, but it is too late to help it, and the potatoes are probably covered with bugs. I will send out to-morrow some Paris green and a couple of blow-guns. There is not much real farm work to do just now, and you will have time for other things. The first and most important thing is to dig a cellar to put your house over; your comfort depends on that. Get the men and horses with plough and scraper out as early as you can to-morrow morning, and hustle. You have nothing to do but dig a big hole seven feet deep inside these lines. I count on you to keep things moving, and I will be out the day after to-morrow."

The mason had finished his estimate, which was $560. After some explanations, I concluded that it was a fair price, and agreed to it, provided the work could be done promptly. The carpenter was not ready to give me figures; he said, however, that he could get a man to move the house for $120, and that he would send me by mail that night an itemized estimate of costs, and also one from a plumber. This seemed like doing a lot of things in one afternoon, so Polly and I started for town content.

"Those people can't be very luxurious out there," said Polly, "but they can have good food and clean beds. They have all out-doors to breathe in, and I do not see what more one can ask on a fine August evening, do you, Mr. Headman?"

I could think of a few things, but I did not mention them, for her first words recalled some scenes of my early life on a backwoods farm: the log cabin, with hardly ten nails in it, the latch-string, the wide-mouthed stone-and-stick chimney, the spring-house with its deep crocks, the smoke-house made of a hollow gum-tree log, the ladder to the loft where I slept, and where the snows would drift on the floor through the rifts in the split clapboards that roofed me over. I wondered if to-day was so much better than yesterday as conditions would warrant us in expecting.

CHAPTER VII

The Horse-and-Buggy Man

August 3 found me at Four Oaks in the early afternoon. A great hollow had been dug for the cellar, and Thompson said that it would take but one more full day to finish it. Piles of material gave evidence that the mason was alert, and the house-mover had already dropped his long timbers, winch, and chains by the side of the farm-house.

While I was discussing matters with Thompson, a smart trap turned into the lot, and a well-set-up young man sprang out of the stylish runabout and said,—

"Dr. Williams, I hear you want more help on your farm."

"I can use another man or two to advantage, if they are good ones."

"Well, I don't want to brag, but I guess I am a good one, all right. I ain't afraid of work, and there isn't much that I can't do on a farm. What wages do you pay?"

I told him my plan of an increasing wage scale, and he did not object. "That includes horse keep, I suppose?" said he.

"I do not know what you mean by 'horse keep.'"

"Why, most of the men on farms around here own a horse and buggy, to use nights, Sundays, and holidays, and we expect the boss to keep the horse. This is my rig. It is about the best in the township; cost me $280 for the outfit."

"See here, young man, this is another specimen of farm economics, and it is one of the worst in the lot. Let me do a small example in mental arithmetic for you. The interest on $280 is $14; the yearly depreciation of your property, without accidents, is at least $40; horse-shoeing and repairs, $20; loss of wages (for no man will keep your horse for less than $4 a month), $48. In addition to this, you will be tempted to spend at least $5 a month more with a horse than without one; that is $60 more. You are throwing away $182 every year without adding $1 to your value as an employee, one ounce of dignity to your employment, or one foot of gain in your social position, no matter from what point you view it.

"Taking it for granted that you receive $25 a month for every month of the year (and this is admitting too much), you waste more than half on that blessed rig, and you can make no provision for the future, for sickness, or for old age. No, I will not keep your horse, nor will I employ any man whose scheme of life doesn't run further than the ownership of a horse and buggy."

"But a fellow must keep up with the procession; he must have some recreation, and all the men around here have rigs."

"Not around Four Oaks. Recreation is all right, but find it in ways less expensive. Read, study, cultivate the best of your kind, plan for the future and save for it, and you will not lack for recreation. Sell your horse and buggy for $200, if you cannot get more, put the money at interest, save $200 out of your wages, and by the end of the year you will be worth over $400 in hard cash and much more in self-respect. You can easily add 1200 a year to your savings, without missing anything worth while; and it will not be long before you can buy a farm, marry a wife, and make an independent position. I will have no horse-and-buggy men on my farm. It's up to you."

"By Jove! I believe you may be right. It looks like a square deal, and I'll play it, if you'll give me time to sell the outfit."

"All right, come when you can. I'll find the work."

That day being Saturday, I told Thompson that I would come out early Monday morning, bringing with me a rough map of the place as I had planned it, and we would go over it with a chain and drive some outlining stakes. I then returned to Exeter, found the carpenter and the plumber, and accepted their estimates,—$630 and $325, respectively. The farm-house moved, finished, furnished, and heated, but not painted or papered, would cost $2630. Painting, papering, window-shades, and odds and ends cost $275, making a total of $2905. It proved a good investment, for it was a comfortable and convenient home for the men and women who afterward occupied it. It has certainly been appreciated by its occupants, and few have left it without regret. We have always tried to make it an object lesson of cleanliness and cheerfulness, and I don't think a man has lived in it for six months without being bettered. It seemed a good deal of money to put on an old farm-house for farm-hands, but it proved one of the best investments at Four Oaks, for it kept the men contented and cheerful workers.

CHAPTER VIII

We Plat the Farm

On Monday I was out by ten o'clock, armed with a surveyor's chain. Thompson had provided a lot of stakes, and we ran the lines, more or less straight, in general accord with my sketch plan. We walked, measured, estimated, and drove stakes until noon. At one o'clock we were at it again, and by four I was fit to drop from fatigue. Farm work was new to me, and I was soft as soft. I had, however, got the general lay of the land, and could, by the help of the plan, talk of its future subdivisions by numerals,—an arrangement that afterward proved definite and convenient. We adjourned to the shade of the big black oak on the knoll, and discussed the work in hand.

"You cannot finish the cellar before to-morrow night," I said, "because it grows slower as it grows deeper; but that will be doing well enough. I want you to start two teams ploughing Wednesday morning, and keep them going every day until the frost stops them. Let Sam take the plough, and have young Thompson follow with the subsoiler. Have them stick to this as a regular diet until I call them off. They are to commence in the wheat stubble where lots six and seven will be. I am going to try alfalfa in that ground, though I am not at all sure that it will do well, and the soil must be fitted as well as possible. After it has had deep ploughing it is to be crossed with the disk harrow; then have it rolled, disk it again, and then use the flat harrow until it feels as near like an ash heap as time will permit. We must get the seed in before September."

"We will need another team if you keep two ploughing and one on the harrow," said Thompson.

"You are right, and that means another $400, but you shall have it. We must not stop the ploughs for anything. Numbers 10, 11, 14, 1, 2, 3, 4, 5, and much of the home lot, ought to be ploughed before snow flies. That means about 160 acres,—80 odd days of steady work for the ploughmen and horses. You will probably find it best to change teams from time to time. A little variety will make it easier for them. As soon as 6 and 7 are finished, turn the ploughs into the 40 acres which make lots 1 to 5. All that must be seeded to pasture grass, for it will be our feeding-ground, and we'll be late with it if we don't look sharp.

"We must have more help, by the way. That horse-and-buggy man, Judson, is almost sure to come, and I will find another. Some of you will have to bunk in the hay for the present, for I am going to send out a woman to help your wife. Six men can do a lot of work, but there is a tremendous lot of work to do. We must fit the ground and plant at least three thousand apple trees before the end of November, and we ought to fence this whole plantation. Speaking of fences reminds me that I must order the cedar posts. Have you any idea how many posts it will take to fence this farm as we have platted it? I suppose not. Well, I can tell you. Twenty-two hundred and fifty at one rod apart, or 1850 at twenty feet apart. These posts must be six feet above and three feet below ground. They will cost eighteen cents each. That item will be $333, for there are seven miles of fence, including the line fence between me and my north neighbor. I am going to build that fence myself, and then I shall know whose fault it is if his stock breaks through. Of course some of the old posts are good, but I don't believe one in twenty is long enough for my purpose."

"What do you buy cedar posts for, when you have enough better ones on the place?" asked Thompson.

"I don't know what you mean."

"Well, down in the wood yonder there's enough dead white oak, standing or on the ground, to make three thousand, nine-foot posts, and one seasoned white oak will outlast two cedars, and it is twice as strong."

"Well, that's good! How much will it cost to get them out?"

"About five cents apiece. A couple of smart fellows can make good wages at that price."

"Good. We will save thirteen cents each. They will cost $93 instead of $333. I don't know everything yet, do I, Thompson?"

"You learn easy, I reckon."

"Keep your eyes and ears open, and if you find any one who can do this job, let him have it, for we are going to be too busy with other things at present. It's time for me to be off. I cannot be out again till

Thursday, for I must find a man, a woman, and a team of horses and all that goes with them. I'll see you on the 8th at any rate."

I was dead tired when I reached home; but there wasn't a grain of depression in my fatigue,—rather a sense of elation. I felt that for the first time in thirty years real things were doing and I was having a hand in them. The fatigue was the same old tire that used to come after a hard day on my father's farm, and the sense was so suggestive of youth that I could not help feeling younger. I have never gotten away from the faith that the real seed of life lies hidden in the soil; that the man who gives it a chance to germinate is a benefactor, and that things done in connection with land are about the only real things. I have grown younger, stronger, happier, with each year of personal contact with the soil. I am thankful for seven years of it, and look forward to twice seven more. I have lost the softness which nearly wilted me that 5th day of August, and with the softness has gone twenty or thirty pounds of useless flesh. I am hard, active, and strong for a man of sixty, and I can do a fair day's work. To tell the truth, I prefer the moderate work that falls to the lot of the Headman, rather than the more strenuous life of the husbandman; but I find an infinite deal to thank the farm for in health and physical comfort.

CHAPTER IX

House-Cleaning

After dinner I telephoned the veterinary surgeon that I wanted another team. He replied that he thought he knew of one that would suit, and that he would let me know the next day. I also telephoned two "want ads." to a morning paper, one for an experienced farm-hand, the other for a woman to do general housework in the country. Polly was to interview the women who applied, and I was to look after the men. That night I slept like a hired man.

Out of the dozen who applied the next day I accepted a Swede by the name of Anderson. He was about thirty, tall, thin, and nervous. He did not fit my idea of a stockman, but he looked like a worker, and as I could furnish the work we soon came to terms.

A few words more about Anderson. He proved a worker indeed. He had an insatiable appetite for work, and never knew when to quit. He was not popular at the farm, for he was too eager in the morning to start and too loath in the evening to stop. His unbridled passion for work was a thing to be deplored, as it kept him thin and nervous. I tried to moderate this propensity, but with no result. Anderson could not be trusted with horses, or, indeed, with animals of any kind, for he made them as nervous as himself; but in all other kinds of work he was the best man ever at Four Oaks. He worked for me nearly three years, and then suddenly gave out from a pain in his left chest and shortness of breath. I called a physician for poor Anderson, and the diagnosis was dilatation of the heart from over-exercise.

"A rare disease among farm-hands, Dr. Williams," said Dr. High, but my conscience did not fully forgive me. I asked Anderson to stay at the farm and see what could be done by rest and care. He declined this, as well as my offer to send him to a hospital. He expressed the liveliest gratitude for kindnesses received and others offered, but he said he must be independent and free. He had nearly $1200 in a savings bank in the city, and he proposed to use it, or such portion of it as was necessary. I saw him two months later. He was better, but not able to work. Hearing nothing from him for three years, a year ago I called at the bank where I knew he had kept his savings. They had sent sums of money to him, once to Rio Janeiro and once to Cape Town. For two years he had not been heard from. Whether he is living or dead I do not know. I only know that a valuable man and a unique farm-hand has disappeared. I never think of Anderson without wishing I had been more severe with him,—more persistent in my efforts to wean him from his real passion. Peace to his ashes, if he be ashes.

That same day I telephoned the Agricultural Implement Company to send me another wagon, with harness and equipment for the team. The veterinary surgeon reported that he had a span of mares for me to look at, but I was too much engaged that day to inspect the team, and promised to do so on the next.

When I reached home, Polly said she had found nothing in the way of a general housework girl for the country. She had seen nine women who wished to do all other kinds of work, but none to fit her wants.

"What do they come for if they don't want the place we described? Do they expect we are to change our plans of life to suit their personal notions?" she asked.

"It's hard to say what they came for or what they want. Their ways are past finding out. We will put in another 'ad.' and perhaps have better luck."

Wednesday, the 7th, I went to see the new team. I found a pair of flea-bitten gray Flemish mares, weighing about twenty-eight hundred pounds. They were four years old, short of leg and long of body, and looked fit. The surgeon passed them sound, and said he considered them well worth the price asked,—$300. I was pleased with the team, and remembered a remark I had heard as a boy from an itinerant Methodist minister at a time when the itinerant minister was supposed to know all there was to know about horse-flesh. This was his remark: "There was never a flea-bitten mare that was a poor horse." In spite of its ambiguity, the saying made an impression from which I never recovered. I always expected great things from flea-bitten grays.

The team, wagon, harness, etc., added $395 to the debit account against the farm. Polly secured her girl,—a green German who had not been long enough in America to despise the country.

"She doesn't know a thing about our ways," said Polly, "but Mrs. Thompson can train her as she likes. If you can spend time enough with green girls, they are apt to grow to your liking."

On Thursday I saw Anderson and the new team safely started for the farm. Then Polly, the new girl, and I took train for the most interesting spot on earth.

Soon after we arrived I lost sight of Polly, who seemed to have business of her own. I found the mason and his men at work on the cellar wall, which was almost to the top of the ground. The house was on wheels, and had made most of its journey. The house mover was in a rage because he had to put the house on a hole instead of on solid ground, as he had expected. "I have sent for every stick of timber and every cobbling block I own, to get this house over that hole; there's no money in this job for me; you ought to have dug the cellar after the house was placed," said he.

I made friends with him by agreeing to pay $30 more for the job. The house was safely placed, and by Saturday night the foundation walls were finished.

Sam and Zeb had made a good beginning on the ploughing, the teams were doing well for green ones, and the men seemed to understand what good ploughing meant. Thompson and Johnson had spent parts of two days in the potato patches in deadly conflict with the bugs.

"We've done for most of them this time," said Thompson, "but we'll have to go over the ground again by Monday."

The next piece of work was to clear the north forty (lots 1 to 5) of all fences, stumps, stones, and rubbish, and all buildings except the cottage. The barn was to be torn down, and the horses were to be temporarily stabled in the old barn on the home lot. Useful timbers and lumber were to be snugly piled, the manure around the barns was to be spread under the old apple trees, which were in lot No. 1, and everything not useful was to be burned. "Make a clean sweep, and leave it as bare as your hand," I told Thompson. "It must be ready for the plough as soon as possible."

Judson, the man with the buggy, reported at noon. He came with bag and baggage, but not with buggy, and said that he came to stay.

"Thompson," said I, "you are to put Judson in charge of the roan team to follow the boys when they are far enough ahead of him. In the meantime he and the team will be with you and Johnson in this house-cleaning. By to-morrow night Anderson and the new team will get in, and they, too, will help on this job. I want you to take personal charge of the gray team,—neither Johnson nor Anderson is the right sort to handle horses. The new team will do the trucking about and the regular farm work, while the other three are kept steadily at the ploughs and harrows."

The cleaning of the north forty proved a long job. Four men and two teams worked hard for ten days, and then it was not finished. By that time the ploughmen had finished 6 and 7, and were ready to begin on No. 1. Judson, with the roans and harrows, was sent to the twenty acres of ploughed ground, and Zeb and his team were put at the cleaning for three days, while Sam ploughed the six acres of old orchard with a *shallow-set* plough. The feeding roots of these trees would have been seriously injured if we had followed the deep ploughing practised in the open. By August 24 about two hundred loads of manure from the barn-yards, the accumulation of years, had been spread under the apple trees, and I felt sure it was well bestowed. Manuring, turning the sod, pruning, and spraying, ought to give a good crop of fruit next year.

We had several days of rain during this time, which interfered somewhat with the work, but the rains were gratefully received. I spent much of my time at Four Oaks, often going every day, and never let more than two days pass without spending some hours on the farm. To many of my friends this seemed a waste of time. They said, "Williams is carrying this fad too far,—spending too much time on it."

Polly did not agree with them, neither did I. Time is precious only as we make it so. To do the wholesome, satisfying thing, without direct or indirect injury to others, is the privilege of every man. To the charge of neglecting my profession I pleaded not guilty, for my profession had dismissed me without so much as saying "By your leave." I was obliged to change my mode of life, and I chose to be a producer rather than a consumer of things produced by others. I was conserving my health, pleasing my wife, and at the same time gratifying a desire which had long possessed me. I have neither apology to make nor regret to record; for as individuals and as a family we have lived healthier, happier, more wholesome, and more natural lives on the farm than we ever did in the city, and that is saying much.

CHAPTER X

Fenced In

On the 26th, when I reached the station at Exeter, I found Thompson and the gray team just starting for the farm with the second load of wire fencing. I had ordered fifty-six rolls of Page's woven wire fence, forty rods in each roll. This fence cost me seventy cents a rod, $224 a mile, or $1568 for the seven miles. Add to this $37 for freight, and the total amounted to $1605 for the wire to fence my land. I got this facer as I climbed to the seat beside Thompson. I did not blink, however, for I had resolved in the beginning to take no account of details until the 31st day of December, and to spend as much on the farm in that time as I could without being wasteful. I did not care much what others thought. I felt that at my age time was precious, and that things must be rushed as rapidly as possible.

I was glad of this slow ride with Thompson, for it gave me an opportunity to study him. I wondered then and afterward why a man of his general intelligence, industry, and special knowledge of the details of farming, should fail of success when working for himself. He knew ten times as much about the business as I did, and yet he had not succeeded in an independent position. Some quality, like broadness of mind or directness of purpose, was lacking, which made him incapable of carrying out a plan, no matter how well conceived. He was like Hooker at Chancellorsville, whose plan of campaign was perfect, whose orders were carried out with exactness, whose army fell into line as he wished, and whose enemy did the obvious thing, yet who failed terribly because the responsibility of the ultimate was greater than he could bear. As second in command, or as corps leader, he was superb; in independent command he was a disastrous failure.

Thompson, then, was a Joe Hooker on a reduced plane,—good only to execute another man's plans. Thompson might have rebutted this by saying that I too might prove a disastrous failure; that as yet I had shown only ability to spend,—perhaps not always wisely. Such rebuttal would have had weight seven years ago, but it would not be accepted to-day, for I have made my campaign and won my battle. The record of the past seven years shows that I can plan and also execute.

Thompson told me that he had found two woodsmen (by scouting around on Sunday) who were glad to take the job of cutting the white-oak posts at five cents each, and that they were even then at work; and that Nos. 6 and 7 would be fitted for alfalfa by the end of the week. He added that the seed ought to be sown as soon thereafter as possible and that a liberal dressing of commercial fertilizer should be sown before the seed was harrowed in.

"I have ordered five tons of fertilizer," I said, "and it ought to be here this week. Sow four bags to the acre."

"Four bags,—eight hundred pounds; that's pretty expensive. Costs, I suppose, $35 to $40 a ton."

"No; $24."

"How's that?"

"Friend at court; factory price; $120 for five tons; $5 freight, making in all $125. We must use at least eight hundred pounds this fall and five hundred in the spring. Alfalfa is an experiment, and we must give it a show."

"Never saw anything done with alfalfa in this region, but they never took no pains with it," said Thompson.

"I hope it will grow for us, for it is great forage if properly managed. The seed will be out this week, and you had best sow it on Monday, the 2d."

"How are you going to seed the north forty?"

"Timothy, red top, and blue grass; heavy seeding, to get rid of the weeds. These lots will all be used as stock lots. Small ones, you think, but we will depend almost entirely upon soiling. I hope to keep a fair sod on these lots, and they will be large enough to give the animals exercise and keep them healthy. I hope the carpenter is pushing things on the house. I want to get you into better quarters as soon as possible, and I want the cottage moved out of the way before we seed the lot."

"They're pushing things all right, I guess; that man Nelson is a hustler."

When I reached the farm I found Johnson and Anderson tearing down the old fence that was our eastern boundary. None of the posts were long enough for my purpose, so all were consigned to the wood-pile.

My neighbor on the north owned just as much land as I did. He inherited it and a moderate bank account from his father, who in turn had it from his. The farm was well kept and productive. The house and barns were substantial and in good repair. The owner did general farming, raised wheat, corn, and oats to sell, milked twenty cows and sent the milk to the creamery, sold one or two cows and a dozen calves each year, and fattened twenty or thirty pigs. He was pretty certain to add a few hundred dollars to his bank account at the end of each season. He kept one man all the time and two in summer. He was a bachelor of twenty-eight, well liked and good to look upon: five feet ten inches in height, broad of shoulder, deep of chest, and a very Hercules in strength. His face was handsome, square-jawed and strong. He was good-natured, but easily roused, and when angry was as fierce as fire. He had the reputation of being the hardest fighter in the country. His name was William Jackson, so he was called Bill. I had met Jackson often, and we had taken kindly to each other. I admired his frank manner and sturdy physique, and he looked upon me as a good-natured tenderfoot, who might be companionable, and who would certainly stir up things in the neighborhood. I went in search of him that afternoon to discuss the line fence, a full mile of which divided our lands.

"I want to put a fence along our line which nothing can get over or under," I said. "I am willing to bear the expense of the new fence if you will take away the old one and plough eight furrows,—four on your land and four on mine,—to be seeded to grass before the wires are stretched. We ought to get rid of the weeds and brush."

"That is a liberal proposition, Dr. Williams, and of course I accept," said Jackson; "but I ought to do more. I'll tell you what I'll do. You are planning to put a ring fence around your land,—three miles in all. I'll plough the whole business and fit it for the seed. I'll take one of my men, four horses, and a grub plough, and do it whenever you are ready."

This settled the fence matter between Jackson and me. The men who cut the posts took the job of setting them, stretching the wire, and hanging the gates, for $400. This included the staples and also the stretching of three strands of barbed wire above the woven wire; two at six-inch intervals on the outside, and one inside, level with the top of the post. Thus my ring fence was six feet high and hard to climb. I have a serious dislike for trespass, from either man or beast, and my boundary fence was made to discourage trespassers. I like to have those who enter my property do so by the ways provided, for "whoso climbeth up any other way, the same is a thief and a robber."

The ring fence was finished by the middle of October. The interior fences were built by my own men during soft weather in winter and spring; and, as I had already paid for the wire and posts, nothing more should be charged to the fence account. In round numbers these seven miles of excellent fence cost me $2100. A lot of money! But the fence is there to-day as serviceable as when it was set, and it will stand for twice seven years more. One hundred dollars a year is not a great price to pay for the security and seclusion which a good fence furnishes. There was no need of putting up so much interior fence. I would save a mile or two if I had it to do again; however, I do not dislike my straight lanes and tightly fenced fields.

CHAPTER XI

The Building Line

Before leaving Four Oaks that day I had a long conversation with Nelson, the carpenter. I had taken his measure, by inquiry and observation, and was willing to put work into his hands as fast as he could attend to it. The first thing was to put him in possession of my plan of a building line.

Two hundred feet south of the north line of the home lot a street or lane was to run due west from the gate on the main road. This was to be the teaming or business entrance to the farm. Commencing three hundred feet from the east end of this drive, the structures were to be as follows: On the south side, first a cold-storage house, then the farm-house, the cottage, the well, and finally the carriage barn for the big house. On the north side of the line, opposite the ice-house, the dairy-house; then a square with a small power-house for its centre, a woodhouse, a horse barn for the farm horses, a granary and a forage barn for its four corners. Beyond this square to the west was the fruit-house and the tool-house—the latter large enough to house all the farm machinery we should ever need. I have a horror of the economy that leaves good tools to sky and clouds without protection. This sketch would not be worked out for a long time, as few of the buildings were needed at once. It was made for the sake of having a general design to be carried out when required; and the water and sewer system had been built with reference to it.

I told Nelson that a barn to shelter the horses was the first thing to build, after the house for the men, and that I saw no reason why two or even three buildings should not be in process of construction at the same time. He said there would be no difficulty in managing that if he could get the men and I could get the money. I promised to do my part, and we went into details.

I wanted a horse barn for ten horses, with shed room for eight wagons in front and a small stable yard in the rear; also a sunken manure vat, ten feet by twenty, with cement walls and floor, the vat to be four feet deep, two feet in the ground and two feet above it. A vat like this has been built near each stable where stock is kept, and I find them perfectly satisfactory. They save the liquid manure, and thus add fifty per cent to the value of the whole. Open sheds protect from sun and rain, and they are emptied as often as is necessary, regardless of season, for I believe that the fields can care for manure better than a compost heap.

I also told Nelson to make plans and estimates for a large forage barn, 75 by 150 feet, 25 feet from floor to rafter plate, with a driving floor through the length of it and mows on either side. A granary, with a capacity of twenty thousand bushels, a large woodhouse, and a small house in the centre of this group where the fifteen horse-power engine could be installed, completed my commissions for that day.

Plans for these structures were submitted in due time, and the work was pushed forward as rapidly as possible. The horse barn made a comfortable home for ten horses, if we should need so many, with food and water close at hand and every convenience for the care of the animals and their harness. The forage barn was not expensive,—it was simply to shelter a large quantity of forage to be drawn upon when needed. The woodhouse was also inexpensive, though large. Wood was to be the principal fuel at Four Oaks, since it would cost nothing, and there must be ample shelter for a large amount. The granary would have to be built well and substantially, but it was not large. The power-house also was a small affair. The whole cost of these five buildings was $8550. The itemized amount is, horse barn, $2000, forage barn, $3400, granary, $2200, woodhouse, $400, power-house, $550.

CHAPTER XII

Carpenters Quit Work

On Friday, August 30, I was obliged to go to a western city on business that would keep me from four to ten days. I turned my face away from the farm with regret. I could hardly realize that I had spent but one month in my new life, the old interests had slipped so far behind. I was reluctant to lose sight, even for a week, of the intensely interesting things that were doing at Four Oaks. Polly said she would go to Four Oaks every day, and keep so watchful an eye on the farm that it could not possibly get away.

"You're getting a little bit maudlin about that farm, Mr. Headman, and it will do you good to get away for a few days. There are *some other* things in life, though I admit they are few, and we are not to forget them. I am up to my ears in plans for the house and the home lot; but I can't quite see what you find so interesting in tearing down old barns and fences and turning over old sods."

"Every heart knoweth its own sorrow, Polly, and I have my troubles."

Friday evening, September 6, I returned from the west. My first greeting was,—

"How's the farm, Polly?"

"It's there, or was yesterday; I think you'll find things running smoothly."

"Have they sowed the alfalfa and cut the oats?"

"Yes."

"Finished the farm-house?"

"No, not quite, but the painters are there, and Nelson has commenced work on two other buildings."

"What time can I breakfast? I must catch the 8.10 train, and spend a long day where things are doing."

Things were humming at Four Oaks when I arrived. Ten carpenters besides Nelson and his son were pounding, sawing, and making confusion in all sorts of ways peculiar to their kind. The ploughmen were busy. Thompson and the other two men were shocking oats. I spent the day roaming around the place, watching the work and building castles. I went to the alfalfa field to see if the seed had sprouted. Disappointed in this, I wandered down to the brook and planned some abridgment of its meanderings. It could be straightened and kept within bounds without great expense if the work were done in a dry season. Polly had asked for a winding brook with a fringe of willows and dogwood, but I would not make this concession to her esthetic taste. This farm land must be useful to the sacrifice of everything else. A winding brook would be all right on the home lot, if it could be found, but not on the farm. A straight ditch for drainage was all that I would permit, and I begrudged even that. No waste land in the cultivated fields, was my motto. I had threshed this out with Polly and she had yielded, after stipulating that I must keep my hands off the home forty.

Over in the woods I found two men at work splitting fence posts. They seemed expert, and I asked them how many they could make in a day.

"From 90 to 125, according to the timber. But we must work hard to make good wages."

"That applies to other things besides post-splitting, doesn't it?"

Closer inspection of the wood lot gratified me exceedingly. Little had been done for it except by Nature, but she had worked with so prodigal a hand that it showed all kinds of possibilities, both for beauty and for utility. Before leaving the place, I had a little talk with Nelson.

"Everything is going on nicely," he said. "I have ten carpenters, and they are a busy lot. If I can only hold them on to the job, things will go well."

"What's the matter? Can't you hold them?"

"I hope so, but there is a hoisters' strike on in the city, and the carpenters threaten to go out in sympathy. I hope it won't reach us, but I'm afraid it will."

"What will you do if the men go out?"

"Do the best I can. I can get two non-union men that I know of. They would like to be on this job now, but these men won't permit it. My son is a full hand, so there will be four of us; but it will be slow work."

"See here, Nelson, I can't have this work slack up. We haven't time. Cold weather will be on before we know it. I'm going to take this bull by the horns. I'll advertise for carpenters in the Sunday papers. Some of those who apply will be non-union men, and I'll hold them over for a few days until we see how the cat jumps. If it comes to the worst, we can get some men to take the place of Thompson and Sam, who are carpenters, and set them at the tools. I will not let this work stop, strike or no strike."

"If you put non-union men on you will have to feed and sleep them on the place. The union will make it hot for them."

"I will take all kinds of care of every man who gives me honest work, you may be sure."

When I returned to town I sent this "ad." to two papers: "Wanted: Ten good carpenters to go to the country." The Sunday papers gave a lurid account of the sentiment of the Carpenters' Union and its sympathetic attitude toward the striking hoisters. The forecast was that there would not be a nail driven if the strike were not settled by Tuesday night. It seemed that I had not moved a day too soon. On Monday thirty-seven carpenters applied at my office. Most of them had union tickets and were not considered. Thirteen, however, were not of the union, and they were investigated. I hired seven on these conditions: wages to begin the next day, Tuesday, and to continue through the week, work or no work. If the strike was ordered, I would take the men to the country and give them steady work until my jobs were finished. They agreed to these conditions, and were requested to report at my office on Wednesday morning to receive two days' pay, and perhaps to be set to work.

I did not go to the farm until Tuesday afternoon. There was no change in the strike, and no reason to expect one. The noon papers said that the Carpenters' Union would declare a sympathetic strike to be on from Wednesday noon.

On reaching Four Oaks I called Nelson aside and told him how the land lay and what I had done.

"I want you to call the men together," said I, "and let me talk to them. I must know just how we stand and how they feel."

Nelson called the men, and I read the reports from two papers on the impending strike order.

"Now, men," said I, "we must look this matter in the face in a businesslike fashion. You have done good work here; your boss is satisfied, and so am I. It would suit us down to the ground if you would continue on until all these jobs are finished. We can give you a lot of work for the best part of the year. You are sure of work and sure of pay if you stay with us. That is all I have to say until you have decided for yourselves what you will do if the strike is ordered."

I left the men for a short time, while they talked things over. It did not take them long to decide.

"We must stand by the union," said the spokesman, "but we'll be damned sorry to quit this job. You see, sir, we can't do any other way. We have to be in the union to get work, and we have to do as the union says or we will be kicked out. It is hard, sir, not to do a hit of a hammer for weeks or months with a family on one's hands and winter coming; but what can a man do? We don't see our way clear in this matter, but we must do as the union says."

"I see how you are fixed," said I, "and I am mighty sorry for you. I am not going to rail against unions, for they may have done some good; but they work a serious wrong to the man with a family, for he cannot follow them without bringing hardships upon his dependent ones. It is not fair to yoke him up with a single man who has no natural claims to satisfy, no mouth to feed except his own; but I will talk business.

"You will be ordered out to-morrow or next day, and you say you will obey the order. You have an undoubted right to do so. A man is not a slave, to be made to work against his will; but, on the other hand, is he not a slave if he is forced to quit against his will? Freedom of action in personal matters is a right which wise men have fought for and for which wise men will always fight. Do you find it in the union? What shall I do when you quit work? How long are you going to stay out? What will become of my interests while you are following the lead of your bell-wethers? Shall my work stop because you have been called out for a holiday? Shall the weeds grow over these walls and my lumber rot while you sit idly by? Not by a long sight! You have a perfect right to quit work, and I have a perfect right to continue.

"The rights which we claim for ourselves we must grant to

others. One man certainly has as defensible a right to work as another man has to be idle. In the legitimate exercise of personal freedom there is no effort at coercion, and in this case there shall be none. If you choose to quit, you will do so without let or hindrance from me; but if you quit, others will take your places without let or hindrance from you. You will be paid in full to-night. When you leave, you must take your tools with you, that there may be no excuse for coming back. When you leave the place, the incident will be closed so far as you and I are concerned, and it will not be opened unless I find some of you trying to interfere with the men I shall engage to take your places. I think you make a serious mistake in following blind leaders who are doing you material injury, for sentimental reasons; but you must decide this for yourselves. If, after sober thought, any of you feel disposed to return, you can get a job if there is a vacancy; but no man who works for me during this strike will be displaced by a striker. You may put that in your pipes and smoke it. Nelson will pay you off to-night."

The strike was ordered for Wednesday. On the morning of that day the seven carpenters whom I had engaged arrived at my office ready for work. I took them to the station and started for Four Oaks. At a station five miles from Exeter we quitted the train, hired two carriages, and were driven to the farm without passing through the village.

We arrived without incident, the men had their dinners, and at one o'clock the hammers and saws were busy again. We had lost but one half day. The two non-union men whom Nelson had spoken of were also at work, and three days later the spokesman of the strikers threw up his card and joined our force. We had no serious trouble. It was thought wise to keep the new men on the place until the excitement had passed, and we had to warn some of the old ones off two or three times, but nothing disagreeable happened, and from that day to this Four Oaks has remained non-unionized.

CHAPTER XIII

Planning for the Trees

The morning of September 17th a small frost fell,—just enough to curl the leaves of the corn and show that it was time for it to be laid by. Thompson, Johnson, Anderson, and the two men from the woods, who were diverted from their post-splitting for the time being, went gayly to the corn fields and attacked the standing grain in the old-fashioned way. This was not economical; but I had no corn reaper, and there was none to hire, for the frost had struck us all at the same time. The five men were kept busy until the two patches—about forty-three acres—were in shock. This brought us to the 24th. In the meantime the men and women moved from the cottage to the more commodious farm-house. Polly had found excuses for spending $100 more on the furnishings of this house,—two beds and a lot of other things. Sunday gave the people a chance to arrange their affairs; and they certainly appreciated their improved surroundings.

The cottage was moved to its place on the line, and the last of the seeding on the north forty was done. Ten tons of fertilizer were sown on this forty-acre tract (at a cost of $250), and it was then left to itself, not to be trampled over by man or beast, except for the stretching of fences or for work around some necessary buildings, until the middle of the following May.

We did not sow any wheat that year,—there was too much else to be done of more importance. There is not much money in wheat-farming unless it be done on a large scale, and I had no wish to raise more than I could feed to advantage. Wheat was to be a change food for my fowls; but just then I had no fowls to feed, and there were more than two hundred bushels in stacks ready for the threshers, which I could hold for future hens.

The ploughmen were now directed to commence deep ploughing on No. 14,—the forty acres set apart for the commercial orchard. This tract of land lay well for the purpose. Its surface was nearly smooth, with a descent to the west and southwest that gave natural drainage. I have been informed that an orchard would do better if the slope were to the northeast. That may be true, but mine has done well enough thus far, and, what is more to the point, I had no land with a northeast slope. The surface soil was thin and somewhat impoverished, but the subsoil was a friable clay in which almost anything would grow if it was properly worked and fed. It was my desire to make this square block of forty acres into a first-class apple orchard for profit. Seven years from planting is almost too soon to decide how well I have succeeded, but the results attained and the promises for the future lead me to believe that there will be no failure in my plan.

The three essentials for beginning such an orchard are: prepare the land properly, get good stock (healthy and true to name), and plant it well. I could do no more this year than to plough deep, smooth the surface, and plant as well as I knew how. Increased fertility must come from future cultivation and top dressing. The thing most prominent in my plan was to get good trees well placed in the ground before cold weather set in. At my time of life I could not afford to wait for another autumn, or even until spring. I had, and still have, the opinion that a fall-planted tree is nearly six months in advance of one planted the following spring. Of course there can be no above-ground growth during that time, but important things are being done below the surface. The roots find time to heal their wounds and to send out small searchers after food, which will be ready for energetic work as soon as the sun begins to warm the soil. The earth settles comfortably about these roots and is moulded to fit them by the autumn rains. If the stem is well braced by a mound of earth, and if a thick mulch is placed around it, much will be done below ground before deep frosts interrupt the work; and if, in the early spring, the mulch and mound are drawn back, the sun's influence will set the roots at work earlier by far than a spring tree could be planted.

Other reasons for fall planting are that the weather is more settled, the ground is more manageable, help is more easily secured, and the nurserymen have more time for filling your order. Any time from October 15 until December 10 will answer in our climate, but early November is the best. I had decided to plant the trees in this orchard twenty-five feet apart each way. In the forty acres there would be fifty-two rows, with fifty-two trees in each row,—or twenty-seven hundred in all. I also decided to have but four varieties of apples in this orchard, and it was important that they should possess a number of virtues. They must come into early bearing, for I was too old to wait patiently for slow-growing trees; they must be of kinds most dependable for yearly crops, for I had no respect for off years; and they must be good enough in color, shape, and quality to tempt the most fastidious market. I studied catalogues and talked with pomologists until my mind was nearly unsettled, and finally decided upon Jonathan, Wealthy, Rome Beauty, and Northwestern Greening,—all winter apples, and all red but the last. I was helped in my decision, so far as the Jonathans and Rome Beauties were concerned, by the discovery that more than half of the old orchard was composed of these varieties.

There is little question as to the wisdom of planting trees of kinds known to have done well in your neighborhood. They are just as likely to do well by you as by your neighbor. If the fruit be to your liking, you can safely plant, for it is no longer an experiment; some one else has broken that ground for you.

In casting about for a reliable nurseryman to whom to trust the very important business of supplying me with young trees, I could not long keep my attention diverted from Rochester, New York. Perhaps the reason was that as a child I had frequently ridden over the plank road from Henrietta to Rochester, and my memory recalled distinctly but three objects on that road,—the house of Frederick Douglass, Mount Hope Cemetery, and a nursery of young trees. Everything else was obscure. I fancy that in fifty years the Douglass house has disappeared, but Mount Hope Cemetery and the tree nursery seem to mock at time. The soil and climate near Rochester are especially favorable to the growing of young trees, and my order went to one of the many reliable firms engaged in this business. The order was for thirty-four hundred trees,—twenty-seven hundred for the forty-acre orchard and seven hundred for the ten acres farthest to the south on the home lot. Polly had consented to this invasion of her domain, for reasons. She said:—

"It is a long way off, rather flat and uninteresting, and I do not see exactly how to treat it. Apple trees are pretty at most times, and picturesque when old. You can put them there, if you will seed the ground and treat it as part of the lawn. I hate your old straight rows, but I suppose you must have them."

"Yes, I guess I shall have to have straight rows, but I will agree to the lawn plan after the third year. You must give me a chance to cultivate the land for three years."

Your tree-man must be absolutely reliable. You have to trust him much and long. Not only do you depend upon him to send you good and healthy stock, but you must trust, for five years at least, that this stock will prove true to name. The most discouraging thing which can befall a horticulturist is to find his new fruit false to purchase labels. After wait, worry, and work he finds that he has not what he expected, and that he must begin over again. It is cold comfort for the tree-man to make good his guarantee to replace all stock found untrue, for five years of irreplaceable time has passed. When you have spent time, hope, and expectation as well as money, looking for results which do not come, your disappointment is out of all proportion to your financial loss, be that never so great. In the best-managed nurseries there will be mistakes, but the better the management the fewer the mistakes. Pay good prices for young trees, and demand the best. There is no economy in cheap stock, and the sooner the farmer or fruit-grower comprehends this fact, the better it will be for him. I ordered trees of three years' growth from the bud,—this would mean four-year-old roots. Perhaps it would have been as well to buy smaller ones (many wise people have told me so), but I was in such a hurry! I wanted to pick apples from these trees at the first possible moment. I argued that a sturdy three-year-old would have an advantage over its neighbor that was only two. However small this advantage, I wanted it in my business—my business being to make a profitable farm in quick time. The ten acres of the home lot were to be planted with three hundred Yellow Transparent, three hundred Duchess of Oldenburg, and one hundred mixed varieties for home use. I selected the Transparent and the Duchess on account of their disposition to bear early, and because they are good sellers in a near market, and because a fruit-wise friend was making money from an eight-year-old orchard of three thousand of these trees, and advised me not to neglect them.

My order called for thirty-four hundred three-year-old apple trees of the highest grade, to be delivered in good condition on the platform at Exeter for the lump sum of $550. The agreement had been made in August, and the trees were to be delivered as near the 20th of October as practicable. Apple trees comprised my entire planting for the autumn of 1895. I wanted to do much other work in that line, but it had to be left for a more convenient season. Hundreds of fruit trees, shade trees, and shrubs have since been planted at Four Oaks, but this first setting of thirty-four hundred apple trees was the most important as well as the most urgent.

The orchard was to be a prominent feature in the factory I was building, and as it would be slower in coming to perfection than any other part, it was wise to start it betimes. I have kicked myself black and blue for neglecting to plant an orchard ten years earlier. If I had done this, and had spent two hours a month in the management of it, it would now be a thing of beauty and an income-producing joy forever,—or, at least, as long as my great-grandchildren will need it.

There is no danger of overdoing orcharding. The demand for fruit increases faster than the supply, and it is only poor quality or bad handling that causes a slack market. If the general farmer will become an expert orchardist, he will find that year by year his ten acres of fruit will give him a larger profit than any forty acres of grain land; but to get this result he must be faithful to his trees. Much of the time they are caring for themselves, and for the owner, too; but there are times when they require sharp attention, and if they do not get it promptly and in the right way, they and the owner will suffer. Fruit growing as a sole occupation requires favorable soil, climate, and market, and also a considerable degree of aptitude on the part of the manager, to make it highly profitable. A fruit-grower in our climate must have other interests if he would make the most of his time. While waiting for his fruit he can raise food for hens and hogs; and if he feeds hens and hogs, he should keep as many cows as he can. He will then use in his own factory all the raw material he can raise. This will again be returned to the land as a by-product, which will not only maintain the fertility of the farm, but even increase it. If his cows are of the best, they will yield butter enough to pay for their food and to give a profit; the skim milk, fed to the hogs and hens, will give eggs and pork out of all proportion to its cost; and everything that grows upon his land can thus be turned off as a finished product for a liberal price, and yet the land will not be depleted. The orchard is better for the hens and hogs and cows, and they are better for the orchard. These industries fit into each other like the folding of hands; they seem mutually dependent, and yet they are often divorced, or, at best, only loosely related. This view may seem to be the result of *post hoc* reasoning, but I think it is not. I believe I imbibed these notions with my mother's milk, for I can remember no time when they were not mine. The psalmist said, "Comfort me with apples"; and the psalmist was reputed a wise man. With only sufficient wisdom to plant an orchard, I live in high expectation of finding the same comfort in my old age.

CHAPTER XIV

Planting of the Trees

September proved as dry as August was wet,—only half an inch of water fell; and the seedings would have been slow to start had they depended for their moisture upon the clouds. By October 1, however, green had taken the place of brown on nearly all the sixty acres we had tilled. The threshers came and threshed the wheat and oats. Of wheat there were 311 bushels, of oats, 1272. We stored this grain in the cottage until the granary should be ready, and stacked the straw until the forage barn could receive it. My plan from the first has been to shelter all forage, even the meanest, and bright oat straw is not low in the scale.

On the 10th the horse stable was far enough advanced to permit the horses to be moved, and the old barn was deserted. A neighbor who had bought this barn at once pulled it down and carted it away. In this transaction I held out several days for $50, but as my neighbor was obdurate I finally accepted his offer. The first entry on the credit side of my farm ledger is, By one old barn, $45. The receipts for October, November, and December, were:—

By one old barn	$45.00
By apples on trees (153 trees at $1.85 each)	283.00
By 480 bushels of potatoes at 30 cents per bushel	144.00
By five old sows, not fat	35.00
One cow	15.00
Three cows	70.00
Two cows	35.00
Three cows, two heifers, nine calves	187.00
Forty-three shoats and gilts, average 162 lb., at 2 cents per lb	139.00
Total	$953.00

The young hogs had eaten most of my small potatoes and some of my corn before we parted with them in late November. These sales were made at the farm, and at low prices, for I was afraid to send such stuff to market lest some one should find out whence it came. The Four Oaks brand was to stand for perfection in the future, and I was not willing to handicap it in the least. Top prices for gilt-edged produce is what intensive farming means; and if there is money in land, it will be found close to this line.

The potatoes had been dug and sold, or stored in the cellar of the farm-house; the apples from the trees reserved for home use had been gathered, and we were ready for the fall planting. While waiting for the stock to arrive, we had time to get in all the hay and most of the straw into the forage barn, which was now under roof.

On Saturday, the 26th, word came that sixteen immense boxes had arrived at Exeter for us. Three teams were sent at once, and each team brought home two boxes. Three trips were made, and the entire prospective orchard was safely landed. Monday saw our whole force at work planting trees. Small stakes had been driven to give the exact centre for each hole, so that the trees, viewed from any direction, would be in straight lines. Sam, Zeb, and Judson were to dig the holes, putting the surface dirt to the right, and the poor earth to the left; I was to prune the roots and keep tab on the labels; Johnson and Anderson were to set the trees,—Anderson using a shovel and Johnson his hands, feet, and eyes; while Thompson was to puddle and distribute the trees. The puddling was easily done. We sawed an oil barrel in halves, placed these halves on a stone boat, filled them two-thirds full of water, and added a lot of fine clay. Into this thin mud the roots of each tree were dipped before planting.

My duty was to shorten the roots that were too long, and to cut away the bruised and broken ones. The top pruning was to be done after the trees were all set and banked. The stock was fine in every respect,—fully up to promise. Watching Johnson set his first tree convinced me that he knew more about planting than I did. He lined and levelled it; he pawed surface dirt into the hole, and churned the roots up and down; more dirt, and he tamped it; still more dirt, and he tramped it; yet more dirt, and he stamped it until the tree stood like a post; then loose dirt, and he left it. I was sure Johnson knew his business too well to need advice from a tenderfoot, so I went back to my root pruning.

We were ten days planting these thirty-four hundred trees, but we did it well, and the days were short. We finished on the 7th of November. The trees were now to be top pruned. I told Johnson to cut every tree in the big orchard back to a three-foot stub, unless there was very good reason for leaving a few inches (never more than six), and I turned my back on him and walked away as I said these cruel words. It seemed a shame to cut these bushy, long-legged, handsome fellows back to dwarfish insignificance and brutish ugliness, but it had to be done. I wanted stocky, thrifty, low-headed business trees, and there was no other way to get them. The trees in the lower, or ten-acre, orchard, were not treated so severely. Their long legs were left, and their bushy tops were only moderately curtailed. We would try both high and low heading.

On the night of November 11 the shredders came and set up their great machine on the floor of the forage barn, ready to commence work the next morning. There were ten men in the shredding gang. I furnished six more, and Bill Jackson came with two others to change work with me; that is, my men were to help him when the machine reached his farm. We worked nineteen men and four teams three and a half days on the forty-three acres of corn, and as a result, had a tremendous mow of shredded corn fodder and an immense pile of half-husked ears. For the use of the machine and the wages of the ten men I paid $105. Poor economy! Before next corn-shredding time I owned a machine,—smaller indeed, but it did the work as well (though not as quickly), and it cost me only $215, and was good for ten years.

The weather had favored me thus far. The wet August had put the ground into good condition for seeding, and the dry September and October had permitted our buildings to be pushed forward, but now everything was to change. A light rain began on the morning of the 15th (I did not permit it to interrupt the shredding, which was finished by noon), and by night it had developed into a steady downpour that continued, with interruptions, for six weeks. November and December of 1895 gave us rain and snow fall equal to twelve and a half inches of water. Plans at Four Oaks had to be modified. There was no more use for the ploughs. Nos. 10 and 11, and much of the home lot were left until spring. I had planned to mulch heavily all the newly set trees, and for this purpose had bought six carloads of manure (at a cost of $72); but this manure could not be hauled across the sodden fields, and must needs be piled in a great heap for use in the spring. The carpenters worked at disadvantage, and the farm men could do little more than keep themselves and the animals comfortable. They did, however, finish one good job between showers. They tile-drained the routes for the two roads on the home lot,—the straight one east and west through the building line, about 1000 feet, and the winding carriage drive to the site of the main house, about 1850 feet. The tile pipe cost $123. They also set a lot of fence posts in the soft ground.

Building progressed slowly during the bad weather, but before the end of December the horse barn, the woodshed, the granary, the forage barn, and the power-house were completed, and most of the machinery was in place. The machinery consisted of a fifteen horse-power engine, with shafting running to the forage barn, the granary, and the woodshed. A power-saw was set in the end of the shed, a grinding mill in the granary, and a fodder-cutter in the forage barn. The cost of these items was:—

Engine and shafting	$187.00
Saw	24.00
Mill	32.00
Feed-cutter and carrier	76.00
Total	$319.00

I gave the services of my two carpenters, Thompson and Sam, during most of this time to Nelson, for I had but little work for them, and he was not making much out of his job.

The last few days of 1895 turned clear and cold, and the barometer set "fair." The change chirked us up, and we ended the year in good spirits.

CHAPTER XV

Polly's Judgment Hall

Before closing the books, we should take account of stock, to see what we had purchased with our money. Imprimis: 320 acres of good land, satisfactory to the eye, well fenced and well groomed; 3400 apple trees, so well planted as to warrant a profitable future; a water and sewer system as good as a city could supply; farm buildings well planned and sufficient for the day; an abundance of food for all stock, and to spare; an intelligent and willing working force; machinery for more than present necessity; eight excellent horses and their belongings; six cows, moderately good; two pigs and two score fowls, to be eaten before spring, and *a lot of fun.* What price I shall have to put against this last item to make the account balance, I can tell better when I foot the other side of the ledger.

But first I must add a few items to the debit account. Moving the cottage cost $30. I paid $134 for grass seed and seed rye. The wage account for six men and two women for five months was $735. Their food account was $277. Of course the farm furnished milk, cream, butter, vegetables, some fruit, fresh pork, poultry, and eggs. There were also some small freight bills, which had not been accounted for, amounting to $31, and $8 had been spent in transportation for the men. Then the farm must be charged with interest on all money advanced, when I had completed my additions. The rate was to be five per cent, and the time three months.

On the last day of the year I went to the farm to pay up to date all accounts. I wished to end the year with a clean score. I did not know what the five months had cost me (I would know that evening), but I did know that I had had "the time of my life" in the spending, and I would not whine. I felt a little nervous when I thought of going over the figures with Polly,—she was such a judicious spender of money. But I knew her criticism would not be severe, for she was hand-in-glove with me in the project. I tried to find fault with myself for wastefulness, but some excellent excuse would always crop up. "Your water tower is unnecessary." "Yes, but it adds to the landscape, and it has its use." "You have put up too much fencing." "True, but I wanted to feel secure, and the old fences were such nests of weeds and rubbish." "You have spent too much money on the farm-house." "I think not, for the laborer is worthy of his hire, and also of all reasonable creature comforts." And thus it went on. I would not acknowledge myself in the wrong; nor, arguing how I might, could I find aught but good in my labors. I devoutly hoped to be able to put the matter in the same light when I stood at the bar in Polly's judgment hall.

The day was clear, cool, and stimulating. A fair fall of snow lay on the ground, clean and wholesome, as country snow always is. I wished that the house was finished (it was not begun), and that the family was with me in it. "Another Christmas time will find us here, God willing, and many a one thereafter."

I spent three hours at the farm, doing a little business and a lot of mooning, and then returned to town. The children were off directly after dinner, intent on holiday festivities, so that Polly and I had the house to ourselves. I felt that we needed it. I invited my partner into the den, lighted a pipe for consolation, unlocked the drawer in which the farm ledger is kept, gave a small deprecatory cough, and said:—

"My dear, I am afraid I have spent an awful lot of money in the last five months. You see there is such a quantity of things to do at once, and they run into no end of money. You know, I—"

"Of course I know it, and I know that you have got the worth of it, too."

Wouldn't that console you! How was I to know that Polly would hail from that quarter? I would have kissed her hand, if she would have permitted such liberty; I kissed her lips, and was ready to defend any sum total which the ledger dare show.

"Do you know how much it is?" said Polly.

"Not within a million!" I was reckless then, and hoped the total would be great, for had not Polly said that she knew I had got the worth of my money? And who was to gainsay her? "It is more than I planned for, I know, but I do not see how I could use less without losing precious time. We started into this thing with the theory that the more we put into it, without waste, the more we would ultimately get out of it. Our theory is just as sound to-day as it was five months ago."

"We will win out all right in the end, Mr. Headman, for we will not put the price-mark on health, freedom, happiness, or fun, until we have seen the debit side of the ledger."

"How much do you want to spend for the house?" said I.

"Do you mean the house alone?"

"No; the house and carriage barn. I'll pay for the trees, shrubs, and kickshaws in the gardens and lawns."

"You started out with a plan for a $10,000 house, didn't you? Well, I don't think that's enough. You ought to give me $15,000 for the house and barn and let me see what I can do with it; and you ought to give it to me right away, so that you cannot spend it for pigs and foolish farm things."

"I'll do it within ten days, Polly; and I won't meddle in your affairs if you will agree to keep within the limit."

"It's a bargain," said Polly, "and the house will be much more livable than this one. What do you think we could sell this one for?"

"About $33,000 or $34,000, I think."

"And will you sell it?"

"Of course, if you don't object."

"Sell, to be sure; it would be foolish to keep it, for we'll be country folk in a year."

"I have a theory," said I, "that when we live on the farm we ought to credit the farm with what it costs us for food and shelter here,—providing, of course, that the farm feeds and shelters us as well."

"It will do it a great deal better. We will have a better house, better food, more company, more leisure, more life, and more everything that counts, than we ever had before."

"We'll fix the value of those things when we've had experience," said I. "Now let's get at the figures. I tell you plainly that I don't know what they foot up,—less than $40,000, I hope."

"Don't let's worry about them, no matter what they say."

This from prudent, provident Polly!

"Certainly not," said I, as bold as a lion.

"There are thirty-five items on the debit side of the ledger and a few little ones on the credit side. Hold your breath while I add them.

"I have spent $44,331 and have received $953, which leaves a debit balance of $43,378."

"That isn't so awfully bad, when you think of all the fun you've had."

"Fun comes high at this time of the year, doesn't it, Polly?"

"Much depends on what you call high. You have waited and worked a long time for this. I won't say a word if you spend all you have in the world. It's yours."

"Mine and yours and the children's; but I won't spend it all. Seventy or seventy-five thousand dollars, besides your house and barn money, shall be my limit. There is still an item of interest to be added to this account.

"Interest! Why, John Williams, do you mean to tell me that you borrowed this money? I thought it was your own to do as you liked with. Have you got to pay interest on it?"

"It was mine, but I loaned it to the farm. Before I made this loan I was getting five per cent on the money. I must now look to the farm for my five per cent. If it cannot pay this interest promptly, I shall add the deferred payment to the principal, and it shall bear interest. This must be done each year until the net income from the farm is greater than the interest account. Whatever is over will then be used to reduce the principal."

"That's a long speech, but I don't think it's very clear. I don't see why a man should pay interest on his own money. The farm is yours, isn't it? You bought it with your own money, didn't you? What difference does it make whether you charge interest or not?"

"Not the least difference in the world to us, Polly, but a great deal to the experiment."

"Oh, yes, I forgot the experiment. And how much interest do you add?"

"Five hundred and forty-two dollars. Also, $75 to the lawyer and $5 for recording the deed, making the whole debt of the farm to me $44,000 even."

"Does it come out just even $44,000? I believe you've manipulated the figures."

"Not on your life! Add them yourself. They were put down at all sorts of times during the past five months. My dear, I wish you a good-night and a happy New Year. You have given me a very happy ending for the old one."

CHAPTER XVI

Winter Work

The new year opened full of all sorts of interests and new projects. There were so many things to plan for and to commence at the farm that we often got a good deal mixed up. I can hardly expect to make a connected narrative of the various plans and events, so will follow each one far enough to launch it and then leave it for future development.

Little snow fell in January and February '96. The weather was average winter weather, and a good deal of outdoor work was done. On the 2d I went to the farm to plan with Thompson an outline for the two months. I had decided to make Thompson the foreman, for I had watched him carefully for five months and was satisfied that I might go farther and fare a great deal worse. Indeed, I thought myself very fortunate to have found such a dependable man. He was temperate and good-natured, and he had a bluff, hearty way with the other men that made it easy for them to accept his directions. He was thorough, too, in his work. He knew how a job should be done, and he was not satisfied until it was finished correctly. He was not a worker for work's sake, as was Anderson, but he was willing to put his shoulder to the wheel for results.

"Wait till I get my shoulder under it," was a favorite expression with him, and I am frank to say that when this conjunction took place there was apt to be something doing. Thompson is still at Four Oaks, and it will be a bad day for the farm when he leaves.

"Thompson," said I, "you are to be working foreman out here, and I want you to put your mind on the business and keep it there. I cannot raise your wages, for I have a system; but you shall have $50 as a Christmas present if things go well. Will you stay on these terms?"

"I will stay, all right, Dr. Williams, and I will give the best I've got. I like the looks of this place, and I want to see how you are going to work it out."

That being settled, I told Thompson of some things that must be done during January and February.

"You must get out a great lot of wood, have it sawed, and store it in the shed, more than enough for a year's use. The wood should be taken from that which is already down. Don't cut any standing trees, even though they are dead. Use all limbs that are large enough, but pile the brushwood where it can be burned. We must do wise forestry in these woods, and we will have an unlimited supply of fuel. I mean that the wood lot shall grow better rather than worse as the years go by. We cannot do much for it now, but more in time. You must see to it that the men are not careless about young trees,—no breaking or knocking down will be in order. Another thing to look after is the ice supply. I will get Nelson to build an ice-house directly, and you must look around for the ice. Have you any idea as to where it can be had?"

"A big company is getting ice on Round Lake three miles west, and I suppose they will sell you what you want," said Thompson, "and our teams can haul it all right."

"What do you suppose they will charge per ton on their platform?"

"From twenty-five to forty cents, I reckon."

"All right, make as good a bargain as you can, and attend to it at the best time. When the teams are not hauling ice or wood, let them draw gravel from French's pit. It will be hard to get it out in the winter, but I guess it can be done, and we will need a lot of it on these roads. Have it dumped at convenient places, and we will put it on the drives in the spring.

"Another thing,—we must have a bridge across the brook on each lane. You will find timbers and planks enough in the piles from the old barns to make good bridges, and the men can do the work. Then there is all that wire for the inside fences to stretch and staple; but mind, no barbed wire is to be put on top of inside fences.

"These five jobs will keep you busy for the next two months, for there'll be only four men besides yourself to do them. I am going to set Sam at the chicken plant. I'll see you before long, and we'll go over the cow and hog plans; but you have your work cut out for the next two months. By the way, how much of an ice-house shall I need?"

"How many cows are you going to milk?"

"About forty when we run at full speed; perhaps half that

number this year."

"Well, then you'd better build a house for four hundred tons. That won't be too big when you are on full time, and it's a mighty bad thing to run short of ice."

I saw Nelson the same day and contracted with him for an ice-house capable of holding four hundred tons, for $900. The walls of the house to be of three thicknesses of lumber with two air spaces (one four inches, the other two) without filling. As a result of the conference with Thompson, I had, before the first of March, a wood-house full of wood, which seemed a supply for two years at full steam; an ice-house nearly full of ice; two serviceable bridges across the brook; the wire fencing almost completed; and eighty loads of gravel,—about one-third of what I needed. The whole cash outlay was,—

300 tons of ice at 30 cents per ton	$90.00
80 tons of gravel at 25 cents per load	20.00
Fence staples	19.00
Total	$129.00

The conference with Sam Jones, the hen man, was deferred until my next visit, and my plans for the cow barn, dairy-house, and hog-house were left to Nelson for consideration, he promising to give me estimates within a few days.

CHAPTER XVII

What Shall We Ask of the Hen?

Sam Jones, the chicken-loving man, was as pleased as a boy with a new top when I began to talk of a hen plant. He had a lot of practical knowledge of the business, for he had *failed* in it twice; and I could furnish any amount of theory, and enough money to prevent disaster.

In his previous attempts he had invested nearly all his small capital in a plant that might yield two hundred eggs a day; he had to buy all foods in small quantities, and therefore at high prices; and he had to give his whole time to a business which was too small and too much on the hand-to-mouth order to give him a living profit. My theory of the business was entirely different. I could plan for results, and, what was more to the point, I could wait for them. Mistakes, accidents, even disasters, were disarmed by a bank account; my bread and butter did not depend upon the temper of a whimsical hen. The food would cost the minimum. All grains and green food, and most of the animal food, in the form of skim milk, would be furnished by the farm. I meant also to develop a plant large enough to warrant the full attention of an able-bodied man. I felt no hesitation about this venture, for I did not intend to ask more of my hens than a well-disposed hen ought to be willing to grant.

I do not ask a hen to lay a double-yolk every day in the year. That is too much to expect of a creature in whom the mother instinct is prominent, and who wishes also to have a new dress for herself at least once in that time. I do not wish a hen to work overtime for me. If she will furnish me with eight dozen of her finished product per annum, I will do the rest. Whatever she does more than that shall redound to her credit. Two-hundred-eggs-a-year hens are scarcer than hens with teeth, and I was not looking for the unusual. A hen can easily lay one hundred eggs in three hundred and sixty-five days, and yet find time for domestic and social affairs. She can feel that she is not a subject for charity, while at the same time she retains her self-respect as a hen of leisure.

I have the highest regard for this domestic fowl, and I would not for a great deal impose a too arduous task upon her. I feel like encouraging her in her peculiar industry, for which she is so eminently fitted, but not like forcing her into strenuous efforts that would rob her of vivacity and dull her social and domestic impulses. No; if the hen will politely present me with one hundred eggs a year, I will thank her and ask no more. Some one will say: "How can you make hens pay if they don't lay more than eight dozen eggs a year? Eggs sometimes sell as low as twelve cents per dozen."

Four Oaks hens never have laid one-cent eggs, and never will. They would quit work if such a price were suggested. Ninety per cent of the eggs from Four Oaks have sold for thirty cents or more per dozen, and the demand is greater than the supply. The Four Oaks certificate that the egg is not thirty-six hours old when it reaches the egg cup, makes two and a half cents look small to those who can afford to pay for the best. To lack confidence in the egg is a serious matter at the breakfast table, and a person who can insure perfect trust will not lack patronage. If, therefore, a hen will lay eight dozen eggs, she is welcome to say to an acquaintance: "I have just handed the Headman a two-dollar bill," for she knows that I have not paid fifty cents for her food.

Of course the wages of the hen man and his food and the interest on the plant must be counted, but I do not propose to count them twice. Four Oaks is a factory where several things are made, each in a measure dependent on, and useful to, the others, and we cannot itemize costs of single products because of this mutual dependence. I feel certain that I could not drop one of the factory's industries without loss to each of the others. For this reason I kept a very simple set of books. I charged the farm with all money spent for it, and credited it with all moneys received. Even now I have no very definite knowledge of what it costs to keep a hen, a hog, or a cow; nor do I care. Such data are greatly influenced by location, method of getting supplies, and market fluctuations. I furnish most of my food, and my own market. My crops have never entirely failed, and I take little heed whether they be large or small. They are not for sale as crops, but as finished products. I am not willing to sell them at any price, for I want them consumed on the place for the sake of the land.

Corn has sold for eighty cents a bushel since I began this experiment, yet at that time I fed as much as ever and was not tempted to sell a bushel, though I could easily have spared five thousand. When it went down to twenty-eight cents, I did not care, for corn and oats to me are simply in transition state,—not commodities to be bought or sold. They cost me, one year with another, about the same. An abundant harvest fills my granaries to overflowing; a bad harvest doesn't deplete them, for I do not sell my surplus for fear that I, too, may have to buy out of a high market. I have bought corn and oats a few times, but only when the price was decidedly below my idea of the feeding value of these grains. I can find more than twenty-eight cents in a bushel of corn, and more than eighteen cents in thirty-two pounds of oats. But I am away off my subject. I began to talk about the hen plant, and have wandered to my favorite fad,—the factory farm.

CHAPTER XVIII

White Wyandottes

"Sam," said I, "I am going to start this poultry plant from just as near the beginning of things as possible. I want you to dispose of every hen on the place within the next twenty days, and to burn everything that has been used in connection with them. We've cleared this land of disease germs, if there were germs in it, by turning it bottom-side up; now let's start free from the pestiferous vermin that make a hen's life unhappy. No stock, either old or young, shall be brought here. When we want to change our breeding, we'll buy eggs from the best fanciers and hatch them in our own incubators. It will then be our own fault if we don't keep our chickens comfortable and free from their enemies. This is sound theory, and we'll try how it works out in practice. Certainly it will be easier to keep clean if we start clean. Not one board or piece of lumber that has been used for any other purpose shall find place in my hen-houses. Eternal vigilance makes a full egg basket; and a full egg basket means a lot of money at the year's end. I will never find fault with you for being too careful Attend to the details in such way as suits you best, provided the result is thorough and everlasting cleanliness. Nothing less will win out, and nothing less will meet the requirements of our factory rules.

"The first thing to do is to get the incubating cellar made. It ought to be four feet in the ground and four feet out of it. Make it ten feet by fifteen, inside measure, and you can easily run five two-hundred-egg incubators. Build it near the south fence in No. 4,—that's the lot for the hens. The walls are to be of brick, and we'll have a brick floor put in, for it's too cold to concrete it now. Gables are to point east and west, and each is to have a window; put the door in the middle of the south wall, and shingle the roof. Digging through three feet of frost will be hard, but it must be done, and done quickly. I want you to start your incubator lamps before the 3d of February."

"I can dig the hole without much trouble,—big fire on the

ground for two or three hours will help,—and I can put on the roof and do all the carpenter work, but I can't lay the brick."

"I'll look out for that part of the job, but I want you to see that things are pushed, for I shall have a thousand eggs here by February 1st and another thousand by the 25th, and these eggs mean money."

"What do you have to pay for them?"

"Ten cents apiece,—$200 for two thousand eggs."

"Well, I should say! Are they hand-painted? I wouldn't have had to quit business if I could have sold my eggs at a quarter of that price."

"That's all right, Sam, but you didn't sell White Wyandotte eggs for hatching. I've contracted with two of the best-known fanciers of Wyandottes in the country to send me five hundred eggs apiece February 1st and 25th. I don't think the price is high for the stock."

"Have you decided to keep 'dottes? I hoped you would try Leghorns; they're great layers."

"Yes, they're great summer layers, but the American birds will beat them hollow in winter; and I must have as steady a supply of eggs as possible. My customers don't stop eating eggs in winter, and they'll be willing to pay more for them at that season. The Leghorn is too small to make a good broiler, and as half the chicks come cockerels, we must look out for that."

"Why do you throw down the Plymouth Rocks? They're bigger than 'dottes, and just as good layers."

"I threw down the barred Plymouth Rocks on account of color; I like white hens best. It was hard to decide between White Rocks and Wyandottes, for there's mighty little difference between them as all-around hens. I really think I chose the 'dottes because the first reply to my letters was from a man who was breeding them."

"They are 'beauts,' all of them, and I'll give them a good chance to spread themselves," said Sam.

"What percentage of hatch may we expect from purchased eggs?"

"About sixty chicks out of every hundred eggs, I reckon."

"That would be doing pretty well, wouldn't it? If we had good luck with the sixty chicks, how many would grow up?"

"Fifty ought to."

"Of these fifty, can we count on twenty-five pullets?"

"Yes."

"That's what I was getting at. You think we might, by good luck, raise twenty-five pullets from each hundred eggs. I'll cut that in the middle and be satisfied with twelve, or even with ten. At that rate the two thousand eggs that cost $200 will give me two hundred pullets to begin the egg-making next November. That's not enough; we ought to raise just twice that number. I'll spend as much more on eggs to be hatched by the middle of April or the first of May, and then we can reasonably expect to go into next winter with four hundred pullets. They will cost the farm a dollar apiece, but the farm will have four hundred cockerels to sell at fifty cents each, which will materially reduce the cost."

"I think you put that pretty low, sir; we ought to raise more than four hundred pullets out of four thousand eggs."

"Everything more will be clear gain. I shall be satisfied with four hundred. We must also get at the brooder house. This is the order in which I want the buildings to stand in the chicken lot: first, the incubating house, 10 feet from the south line; 40 feet north of this, the brooder house; and 120 feet north of that, the first hen-house, with runs 100 feet deep. We'll build other houses for the birds as we need them. They are all to face to the south. If the brooder house is 50 feet long and 15 feet wide, it can easily care for the eight hundred chicks, and for half as many more, if we are lucky enough to get them.

"We'll have a five-foot walk against the north wall of this house, and a ten-foot space north and south through the centre for heating plant and food. This will leave a space at each side ten by twenty feet, to be cut into five pens four feet by ten, each of which will mother a hundred chicks or more. There must be plenty of glass in the south wall, and we'll use overhead water pipes in each hover.

"There's no hurry about the poultry-houses. You can build one in the early summer, and perhaps another in the fall. I expect you to do the carpenter work on these houses. I'll see the mason at once and have him ready by the time you've dug the hole. The incubators will be here in good time, and we want everything ready for work as soon as the eggs arrive."

Sam was pleased with his job; it was exactly to his liking. He took real delight in caring for fowls, and he was especially anxious to prove to me that it was not so much lack of knowledge as lack of capital that had caused the downfall of his previous efforts. Sam could not then understand why one man could sell his eggs at thirty-six cents a dozen when his neighbor could get only sixteen; he found out later.

The mason's work for the incubator house and the foundation wall for the brooder house cost $290. The lumber bill for these two, including doors and windows, was $464. The five incubators, $65, and the hot-water heater for the brooder house, $68, made the total $897. Add to this $400 paid during two months for eggs, and we have $1297 as the cost of starting the poultry plant.

CHAPTER XIX

Fried Pork

I had given Nelson this sketch as a guide in working out the plan for the cow barn: Length over all, 130 feet; width, 40 feet. This parallelogram was to be divided lengthwise into three equal spaces, one in the centre for a driveway, and one on each side for the cow platforms and feeding mangers. Twenty feet at the west end of the barn was partitioned off, one corner for a small granary, the other for a kitchen in which the food was to be prepared. These rooms were each thirteen feet by twenty. At the other end of the building, ten feet on each side was given over to hospital purposes,—a lying-in ward ten feet by thirteen being on each side of the driveway.

The foundation for this building was to be of stone, and the entire floor of cement; and the walls were to be sealed within and sheeted without, and then covered with ship lap boards, making three thicknesses of boards. It was to be one story high. An east-and-west passage, cutting the main drive at right angles, divided the barn at its middle. At the south end of this passage was a door leading to the dairy-house, which was on the building line 150 feet away. The four spaces made by these passages were each subdivided into ten stalls five feet wide. Two doors on the north and two on the south gave exit for the cows. I had placed my limit at forty milch cows, and I thought this stable would furnish suitable quarters for that number. If I had to rebuild, I would make some modifications. Experience is a good teacher; but the stable has served its purpose, and I cannot quarrel with the results. The chief defect is in the distribution of water. The supply is abundant, but it is let on only in the kitchen, whence it is supplied to the cows by means of a hose or a barrel swung between wheels.

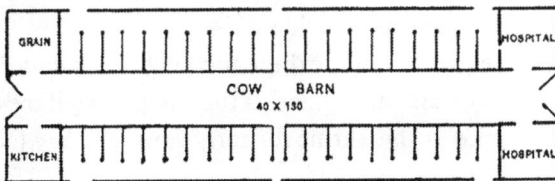

GRAIN · HOSPITAL · COW BARN 40 X 130 · KITCHEN · HOSPITAL

In the kitchen are appliances for mixing and cooking food, and for warming the drinking water in winter. Nelson and I discussed the sketch plan given below, and he found some fault with it. I would not be dissuaded from my views, however, and Nelson had to yield. I was as opinionated in those days as a theoretical amateur is apt to be; and it was hard to give up my theories at the suggestion of a person who had only experience to guide him. The best plan, as I have long since learned, is to mix the two and use the solid substance that results from their combination.

We located the site of the building, and talked plans until the low sun of January 8th disappeared in the west. Then we adjourned to the sitting room of the farm-house to finish the matter so far as was possible. An hour and a half passed, and we were in fair accord, when Mrs. Thompson came into the room to say that supper was ready, and to ask us to join the men at table before starting homeward. I was glad of the opportunity, for I was curious to know if Mrs. Thompson set a good table. We went into the dining room just as the farm family was ready to sit down. There were ten of us,—two women, six men, Nelson, and myself; and as we sat down, I noticed with pleasure that each had evidently taken some thought of the obligations which a table ought to impose. The table was clothed in clean white, and there was a napkin at each plate. Nelson and I had the only perfectly fresh ones, and this I took as evidence that napkins were usual. The food was all on the table, and was very satisfactory to look at. Thompson sat at one end, and before him, on a great platter, lay two dozen or more pieces of fried salt pork, crisp in their shells of browned flour, and fit for a king. On one side of the platter was a heaping dish of steaming potatoes. A knife had been drawn once around each, just to give it a chance to expand and show mealy white between the gaping circles that covered its bulk. At the other side was a boat of milk gravy, which had followed the pork into the frying-pan and had come forth fit company for the boiled potatoes. I went back forty years at one jump, and said,—

"I now renew my youth. Is there anything better under the sun than fried salt pork and milk gravy? If there is, don't tell me of it, for I have worshipped at this shrine for forty years, and my faith must not be shaken."

Such a supper twice or thrice a week would warm the cockles of my old heart; but Polly says, "No modern cook can make these things just right; and if not just right, they are horrid." That is true; it takes an artist or a mother to fry salt pork and make milk gravy.

There were other things on the table,—quantities of bread and butter, apple sauce (in a dish that would hold half a peck), stacks of fresh ginger-bread, tea, and great pitchers of milk; but naught could distract my attention from the *pièce de résistance*. Thrice I sent my plate back, and then could do no more. That meal convinced me that I could trust Mrs. Thompson. A woman who could fry salt pork as my mother did, was a woman to be treasured.

I left the farm-house at 7, and reached home by 8.45. Polly was not quite pleased with my late hours; she said it did not worry her not to know where I was, but it was annoying.

"Can't you have a telephone put into the farm-house? It would be convenient in a lot of ways."

"Why, of course; I don't see why it can't be done at once. I'll make application this very night."

It was six weeks before we really got a wire to the farm, but after that we wondered how we ever got along without it.

CHAPTER XX

A Ration for Product

Nelson was to commence work on the cow-house at once; at least, the mason was. I left the job as a whole to Nelson, and he made some sort of contract with the mason. The agreement was that I should pay $4260 for the barn complete. The machinery we put into it was very simple,—a water heater and two cauldrons for cooking food. All three cost about $60.

Thompson had selected six cows, from those bought with the place, as worth wintering. They were now giving from six to eight quarts each, and were due to come in in April and May. An eight-quart-a-day cow was not much to my liking, but Thompson said that with good care they would do better in the spring. "Four of those cows ought to make fine milkers," he said; "they are built for it,—long bodies, big bags, milk veins that stand out like crooked welts, light shoulders, slender necks, and lean heads. They are young, too; and if you'll dehorn them, I believe they'll make your thoroughbreds hump themselves to keep up with them at the milk pail. You see, these cows never had more than half a chance to show what they could do. They have never been 'fed for milk.' Farmers don't do that much. They think that if a cow doesn't bawl for food or drink she has enough. I suppose she has enough to keep her from starving, and perhaps enough to hold her in fair condition, but not enough to do this and fill the milk pail, too. I read somewhere about a ration for 'maintenance' and one for 'product,' and there was a deal of difference. Most farmers don't pay much attention to these things, and I guess that's one reason why they don't get on faster."

"You've got the whole matter down fine in that 'ration for product,' Thompson, and that's what we want on this farm. A ration that will simply keep a cow or a hen in good health leaves no margin for profit. Cows and hens are machines, and we must treat them as such. Crowd in the raw material, and you may look for large results in finished product. The question ought always to be, How much can a cow eat and drink? not, How little can she get on with? Grain and forage are to be turned into milk, and the more of these foods our cows eat, the better we like it. If these machines work imperfectly, we must get rid of them at once and at any price. It will not pay to keep a cow that persistently falls below a high standard. We waste time on her, and the smooth running of the factory is interrupted. I'm going to place a standard on this farm of nine thousand pounds a year for each matured cow; I don't think that too high. If a cow falls much below that amount, she must give place to a better one, for I'm not making this experiment entirely for my health. The standard isn't too high, yet it's enough to give a fine profit. It means at least three hundred and fifty pounds of butter a year, and in this case the butter means at least thirty cents a pound, or more than $100 a year for each cow. This is all profit, if one wishes to figure it by itself, for the skimmed milk will more than pay for the food and care. But why did you say dehorn the cows?"

"Well, I notice that a man with a club is almost sure to find some use for it. If he isn't pounding the fence or throwing it at a dog, he's snipping daisies or knocking the heads off bull-thistles. He's always doing something with it just because he has it in his hand. It's the same way with a cow. If she has horns, she'll use them in some way, and they take her mind off her business. No, sir; a cow will do a lot better without horns. There's mighty little to distract her attention when her clubs are gone."

"What breeds of cows have you handled, Thompson?"

"Not any thoroughbreds that I know of; mostly common kinds and grade Jerseys or Holsteins."

"I'm going to put a small herd of thorough bred Holsteins on the place."

"Why don't you try thoroughbred Jerseys' They'll give as much butter, and they won't eat more than half as much."

"You don't quite catch my idea, Thompson. I want the cow that will eat the most, if she is, at the same time, willing to pay for her food. I mean to raise a lot of food, and I want a home market for it. What comes from the land must go back to it, or it will grow thin. The Holstein will eat more than the Jersey, and, while she may not make more butter, she will give twice as much skimmed milk and furnish more fertilizer to return to the land. Fresh skimmed milk is a food greatly to be prized by the factory-farm man; and when we run at full speed, we shall have three hundred thousand pounds of it to feed.

"I have purchased twenty three-year-old Holstein cows, in calf to advanced registry bulls, and they are to be delivered to me March 10. I shall want you to go and fetch them. I also bought a young bull from the same herd, but not from the same breeding. These twenty-one animals will cost, by the time they get here, $2200. I shall give the bull to my neighbor Jackson. He will be proud to have it, and I shall be relieved of the care of it. Be good to your neighbor, Thompson, if by so doing you can increase the effectiveness of the factory farm. We will start the dairy with twenty thoroughbreds and six scrubs. I shall probably buy and sell from time to time; but of one thing I am certain: if a cow cannot make our standard, she goes to the butcher, be she mongrel or thoroughbred. What do you think of Judson as a probable dairyman?"

"I shouldn't wonder if he would do first-rate. He's a quiet fellow, and cows like that. He has those roans tagging him all over the place; and if a horse likes a man, it's because he's nice and quiet in his ways. I notice that he can milk a cow quicker than the other men, and it ain't because he don't milk dry—I sneaked after him twice. The cow just gives down for him better than for the others."

CHAPTER XXI

The Razorback

We have now launched three of the four principal industries of our factory farm. The fourth is perhaps the most important of all, if a single member of a group of mutually dependent industries can have this distinction. There is no question that the farmer's best friend is the hog. He will do more for him and ask less of him than any other animal. All he asks is to be born. That is enough for this non-ruminant quadruped, who can find his living in the earth, the roadside ditch, or the forest, and who, out of a supply of grass, roots, or mast, can furnish ham and bacon to the king's taste and the poor man's maintenance. The half-wild razorback, with never a clutch of corn to his back, gives abundant food to the mountaineer over whose forest he ranges. The cropped or slit ear is the only evidence of human care or human ownership. He lives the life of a wild beast, and in the autumn he dies the death of a wild beast; while his flesh, made rich with juices of acorns, beechnuts, and other sweet masts, nourishes a man whose only exercise of ownership is slaughter. The hog that can make his own living, run like a deer, and drink out of a jug, has done more for the pioneer and the backwoodsman than any other animal.

Take this semi-wild beast away from his wild haunts, give him food and care, and he will double his gifts. Add a hundred generations of careful selection, until his form is so changed that it is beyond recognition, and again the product will be doubled. The spirit of swine is not changed by civilization or good breeding; such as it was on that day when the herd "ran down a steep place and was drowned in the sea," such it is to-day. A fixed determination to have its own way dominated the creature then, and a pig-headed desire to be the greatest food-producing machine in the world is its ruling passion now. That the hog has succeeded in this is beyond question; for no other food animal can increase its own weight one hundred and fifty fold in the first eight months of its life.

All over the world there is a growing fondness for swine flesh, and the ever increasing supply doesn't outrun the demand. Since the dispersion of the tribes of Israel there has been no persistent effort to depopularize this wonderful food maker. Pig has more often been the food of the poor than of the rich, but now rich and poor alike do it honor. Old Ben Jonson said:—

"Now pig is meat, and a meat that is nourishing and may be desired, and consequently eaten: it may be eaten; yea, very exceedingly well eaten."

Hundreds have praised the rasher of ham, and thousands the flitch of bacon; it took the stroke of but one pen to make roast pig classical.

The pig of to-day is so unlike his distant progenitor that he would not be recognized; if by any chance he were recognized, it would be only with a grunt of scorn for his unwieldy shape and his unenterprising spirit. Gone are the fleet legs, great head, bulky snout, terrible jaws, warlike tusks, open nostrils, flapping ears, gaunt flanks, and racing sides; and with these has gone everything that told of strength, freedom, and wild life. In their place has come a cuboidal mass, twice as long as it is broad or high, with a place in front for mouth and eyes, and a foolish-looking leg under each corner. A mighty fall from "freedom's lofty heights," but a wonderfully improved machine. The modern hog is to his progenitor as the man with the steam-hammer to the man with the stone-hammer,—infinitely more useful, though not so free.

It is not easy to overestimate the value of swine to the general farmer; but to the factory farmer they are indispensable. They furnish a profitable market for much that could not be sold, and they turn this waste material into a surprising lot of money in a marvellously short time. A pig should reach his market before he is nine months old. From the time he is new-born until he is 250 days old, he should gain at least one pound a day, which means five cents, in ordinary times. During this time he has eaten, of things which might possibly have been sold, perhaps five dollars' worth. At 250 days, with a gain of one pound a day, he is worth, one year with another, $12.50. This is putting it too low for my market, but it gives a profit of not less than $6 a head after paying freight and commissions. It is, then, only a question of how many to keep and how to keep them. To answer the first half of this question I would say, Keep just as many as you can keep well. It never pays to keep stock on half rations of food or care, and pigs are not exceptions. In answering the other half of the question, how to keep them, I shall have to go into details of the first building of a piggery at Four Oaks.

As in the case of the hens, I determined to start clean. Hogs had been kept on the farm for years, and, so far as I could learn, there had been no epizoötic disease. The swine had had free range most of the time, and the specimens which I bought were healthy and as well grown as could be expected. They were not what I wanted, either in breed or in development, so they had been disposed of, all but two. These I now consigned to the tender care of the butcher, and ordered the sty in which they had been kept to be burned.

I had planned to devote lot No. 2 to a piggery. There are five acres in this lot, and I thought it large enough to keep four or five hundred pigs of all sizes in good health and good condition for forcing. Some of the swine, not intended for market, would have more liberty; but close confinement in clean pens and small runs was to be the rule. To crowd hogs in this way, and at the same time to keep them free from disease, would require special vigilance. The ordinary diseases that come from damp and draughts could be fended off by carefully constructed buildings. Cleanliness and wholesome food ought to do much, and isolation should accomplish the rest. I have established a perfect quarantine about my hog lot, and it has never been broken. After the first invoices of swine in the winter and spring of 1896, no hog, young or old, has entered my piggery, save by the way of a sixty-day quarantine in the wood lot, and very few by that way.

My pigs are several hundred yards from the public roads, and my neighbor, Jackson, has planted a young orchard on his land to the north of my hog lots, and permits no hogs in this planting. I have thus secured practical isolation. I have rarely sent swine to fairs or stock shows. In the few instances in which I have broken this rule I have sold the stock shown, never returning it to Four Oaks.

Isolation, cleanliness, good food, good water, and a constant supply of ashes, charcoal, and salt, have kept my herd (thus far) from those dreadfully fatal diseases that destroy so many swine. If I can keep the specific micro-organism that causes hog-cholera off my place, I need not fear the disease. The same is true of swine plague. These diseases are of bacterial origin, and are communicated by the transference of bacteria from the infected to the non-infected. I propose to keep my healthy herd as far removed as possible from all sources of infection. I have carried these precautions so far that I am often scoffed at. I require my swineherd, when returning from a fair or a stock show, to take a full bath and to disinfect his clothing before stepping into the pig-house. This may seem an unnecessary refinement in precautionary measures, but I do not think so. It has served me well: no case of cholera or plague has shown itself at Four Oaks.

What would I do if disease should appear? I do not know. I think, however, that I should fight it as hard as possible at close quarters, killing the seriously ill, and burning all bodies. After the scourge had passed I would dispose of all stock as best I could, and then burn the entire plant (fences and all), plough deep, cover the land white as snow with lime, leave it until spring, plough again, and sow to oats. During the following summer I would rebuild my plant and start afresh. A whole year would be lost, and some good buildings, but I think it would pay in the end. There would be no safety for the herd while a single colony of cholera or plague bacteria was harbored on the place; and while neither might, for years, appear in virulent form, yet there would be constant small losses and constant anxiety. One cannot afford either of these annoyances, and it is usually wise to take radical measures. If we apply sound business rules to farm management, we shall at least deserve success.

I chose to keep thoroughbred swine for the reason that all the standard varieties are reasonably certain to breed true to a type which, in each breed, is as near pork-making perfection as the widest experience can make it. Most of our good hogs are bred from English or Chinese stock. Modifications by climate, care, crossing, and wise selection have procured a number of excellent varieties, which are distinct enough to warrant separate names, but which are nearly equal as pork-makers.

In color one could choose between black, black and white, and white and red. I wanted white swine; not because they are better than swine of other colors, for I do not think they are, but for æsthetic reasons. My poultry was to be white, and white predominated in my cows; why should not my swine be white also,—or as white as their habits would permit? I am told on all sides that the black hog is the hardiest, that it fattens easier, and that for these reasons it is a better all-round hog. This may be true, but I am content with my white ones. When some neighbor takes a better bunch of hogs to market, or gets a better price for them, than I do, I may be persuaded to think as he talks. Thus far I have sold close to the top of the market, and my hogs are never left over.

Perhaps my hogs eat more than those of my neighbors. I hope they do, for they weigh more, on a "weight for age" scale, and I do not think they are "air crammed," for "you cannot fatten capons so." I am more than satisfied with my Chester Whites. They have given me a fine profit each year, and I should be ungrateful if I did not speak them fair.

I wished to get the hog industry started on a liberal scale, and scoured the country, by letter, for the necessary animals. I found it difficult to get just what I wanted. Perhaps I wanted too much. This is what I asked for: A registered young sow due to farrow her second litter in March or April. By dint of much correspondence and a considerable outlay of money, I finally secured nineteen animals that answered the requirements. I got them in twos and threes from scattered sources, and they cost an average price of $31 per head delivered at Four Oaks. A young boar, bred in the purple, cost $27. My foundation herd of Chester Whites thus cost me $614,—too much for an economical start; but, again, I was in a hurry.

The hogs began to arrive in February, and were put into temporary quarters pending the building of the house for the brood sows, which house must now be described.

It was a low building, 150 by 30 feet, divided by a six-foot alley-way into halves, each 150 by 12 feet. Each of these halves was again divided into fifteen pens 10 by 12 feet, with a 10 by 30 run for each pen. This was the general plan for the brood-house for thirty sows. At the east end of this house was a room 15 by 30 feet for cooking food and storing supplies for a few days. The building was of wood with plank floors. It stands there yet, and has answered its purpose; but it was never quite satisfactory. I wanted cement floors and a more sightly building. I shall probably replace it next year. When it was built the weather was unfavorable for laying cement, and I did not wish to wait for a more clement season. The house and the fences for the runs cost $2100.

On the 6th of March Thompson called me to one of the temporary pens and showed me a family of the prettiest new-born animals in the world,—a fine litter of no less than nine new-farrowed pigs. I felt that the fourth industry was fairly launched, and that we could now work and wait.

CHAPTER XXII

The Old Orchard

March was unusually raw even for that uncooked month. The sun had to cross the line before it could make much impression on the deep frost. After the 15th, however, we began to find evidences that things were stirring below ground. The red and yellow willows took on brighter colors, the bark of the dogwood assumed a higher tone, and the catkins and lilac buds began to swell with the pride of new sap.

If our old orchard was to be pruned while dormant, it must be done at once. Thompson and I spent five days of hard work among the trees, cutting out all dead limbs, crossing branches, and suckers. We called the orchard old, but it was so only by comparison, for it was not out of its teens; and I did not wish to deal harshly with it. A good many unusual things were being done for it in a short time, and it was not wise to carry any one of them too far. It had been fertilized and ploughed in the fall, and now it was to be pruned and sprayed,—all innovations. The trees were well grown and thrifty. They had given a fair crop of fruit last year, and they were well worth considerable attention. They could not hereafter be cultivated, for they were all in the soiling lot for the cows, but they could be pruned and sprayed. The lack of cultivation would be compensated by the fertilization incident to a feeding lot. The trees would give shade and comfort to the cows, while the cows fed and nourished the trees,—a fair exchange.

The crop of the year before, though half the apples were stung, had brought nearly $300. With better care, and consequently better fruit, we could count on still better results, for the varieties were excellent (Baldwins, Jonathans, and Rome Beauties); so we trimmed carefully and burned the rubbish. This precaution, especially in the case of dead limbs, is important, for most dead wood in young trees is due to disease, often infectious, and should be burned at once.

I bought a spraying-pump (for $13), which was fitted to a sound oil barrel, and we were ready to make the first attack on fungus disease with the Bordeaux mixture. This was done by Johnson and Anderson late in the month. Another vigorous spraying with the same mixture when the buds were swelling, another when the flower petals were falling, and still another when the fruit was as large as peas (the last two sprayings had Paris green added to the Bordeaux mixture), and the fight against apple enemies was ended for that year.

Thompson had gone for the cows. He left March 9, and returned with the beauties on Friday the 17th. They were all my fancy had painted them,—large, gentle-eyed, with black and white hair over soft butter-yellow skin, and all the points that distinguish these marvellous milk-machines. They were bestowed as needs must until the cow barn was completed. One of them had dropped a bull calf two days before leaving the home farm. The calf had been left, and the mother was in an uncomfortable condition, with a greatly distended udder and milk streaming from her four teats, though Thompson had relieved her thrice while *en route*.

I was greatly pleased with the cows, but must not spend time on them now, for things are happening in my factory faster than I can tell of them. Johnson had built some primitive hotbeds for early vegetables out of old lumber and oiled muslin. He had filled them with refuse from the horse stable and had sown his seeds.

CHAPTER XXIII

The First Hatch

On February 3 the incubator lamps were lighted under the first invoice of one thousand eggs. The incubating cellar was to Sam's liking, and he felt confident that three weeks of strict attention to temperature, moisture, and the turning of eggs, would bring results beyond my expectations.

After the seventh day, on which he had tested or candled the eggs, he was willing to promise almost anything in the way of a hatch, up to seventy-five or eighty per cent. In the intervals of attendance on the incubators he was hard at work on the brooder-house, which must be ready for its first occupants by the 25th. Everything went smoothly until the 18th. That morning Sam met me with a long face.

"Something went wrong with one of my lamps last night," said he. "I looked at them at ten o'clock and they were all right, but at six this morning one of the thermometers was registering 122°, and the whole batch was cooked."

"Not the whole thousand, Sam!"

"No, but 170 fertile eggs, and that spoils a twenty-dollar bill and a lot of good time. What in the name of the black man ever got into that lamp of mine is more than I know. It's just my luck!"

"It's everybody's luck who tries to raise chickens by wholesale, and we must copper it. Don't be downed by the first accident, Sam; keep fighting and you'll win out."

The brooder-house was ready when the first chicks picked the shells on the 24th, and within thirty-six hours we had 503 little white balls of fluff to transfer from the four incubators to the brooder-house. We put about a hundred together in each of five brooders, fed them cut oats and wheat with a little coarse corn meal and all the fresh milk they could drink, and they throve mightily.

The incubators were filled again on the 26th, and from that hatch we got 552 chicks. On the 21st of March they were again filled, and on the 13th of April we had 477 more to add to the colony in the brooder-house. For the last time we started the lamps April 15th, and on the 6th of May we closed the incubating cellar and found that 2109 chicks had been hatched from the 4000 eggs. The last hatch was the best of all, giving 607. I don't think we have ever had as good results since, though to tell the truth I have not attempted to keep an exact count of eggs incubated. My opinion is that fifty per cent is a very good average hatch, and that one should not expect more.

In September, when the young birds were separated, the census report was 723 pullets and 764 cockerels, showing an infant mortality of 622, or twenty-nine per cent. The accidents and vicissitudes of early chickenhood are serious matters to the unmothered chick, and they must not be overlooked by the breeder who figures his profits on paper.

After the first year I kept no tabs on the chickens hatched; my desire was to add each year 600 pullets to my flock, and after the third season to dispose of as many hens. It doesn't pay to keep hens that are more than two and a half years old. I have kept from 1200 to 1600 laying hens for the past six years. I do not know what it costs to feed one or all of them, but I do know what moneys I have received for eggs, young cockerels, and old hens, and I am satisfied.

There is a big profit in keeping hens for eggs if the conditions are right and the industry is followed, in a businesslike way, in connection with other lines of business; that is, in a factory farm. If one had to devote his whole time to the care of his plant, and were obliged to buy almost every morsel of food which the fowls ate, and if his market were distant and not of the best, I doubt of great success; but with food at the lowest and product at the highest, you cannot help making good money. I do not think I have paid for food used for my fowls in any one year more than $500; grits, shells, meat meal, and oil meal will cover the list. I do not wish to induce any man or woman to enter this business on account of the glowing statements which these pages contain. I am ideally situated. I am near one of the best markets for fine food; I can sell all the eggs my hens will lay at high prices; food costs the minimum, for it comes from my own farm; I utilize skim-milk, the by-product from another profitable industry, to great advantage; and I had enough money to carry me safely to the time of product. In other words, I could build my factory before I needed to look to it for revenue. I do not claim that this is the only way, but I do claim that it is the way for the fore-handed middle-aged man who wishes to change from city to country life without financial loss. Younger people with less means can accomplish the same results, but they must offset money by time. The principle of the factory farm will hold as well with the one as with the other.

To intensify farming is the only way to get the fat of the land. The nations of the old world have nearly reached their limit in food production. They are purchasers in the open market. This country must be that market; and it behooves us to look to it that the market be well stocked. There is land enough now and to spare, but will it be so fifty or a hundred years hence? Our arid lands will be made fertile by irrigation, but they will add only a small percentage to the amount already in quasi-cultivation. Our future food supplies must be drawn largely from the six million farms now under fences. These farms must be made to yield fourfold their present product, or they will fall short, not only of the demands made upon them, but also of their possibilities. That is why I preach the gospel of intensive farming, for grain, hay, market, and factory farm alike.

I will put the chickens out of the way for the present, referring to them from time to time and indicating their general management, the cost of their houses and food, and the amount of money received for eggs and fowls. I do not think my plant would win the approval of fanciers, and it is not in all ways up to date; but it is clean, healthy, and commodious, and the birds attend as strictly to business as a reasonable owner could wish. I shall be glad to show it to any one interested enough to search it out, and to go into the details of the business and show how I have been able to make it so remunerative.

Sam is with me no longer. For three years he did good service and saved money, and the lurid nose grew dim. There is, however, a limit to human endurance. Like victims of other forms of circular insanity, the dipsomaniac completes his cycle in an uncertain period and falls upon bad times. For a month before we parted company I saw signs of relapse in Sam. He was loquacious at times, at other times morose. He talked about going into business for himself, and his nose took on new color. I labored with him, but to no purpose; the spirit of unrest was upon him, and it had to work its own. I held him firm long enough to secure another man, and then we parted, he to do business for himself, I to get on as best I could. Sam painted his nose and raised chickens and other things until his savings had flown; then he got a position with a woman who runs a broiler plant, and for two years he has given good service. He will probably continue in ways of well-doing until the next cycle is complete, when the beacon light will blaze afresh and he will follow it on to the rocks. Such a man is more to be pitied than condemned, for his anchor is sure to drag at times.

CHAPTER XXIV

The Holstein Milk Machine

During the month of March the teams hauled more gravel. They also distributed the manure that had been purchased in the fall for mulching the trees. While the ground was still frozen this mulch was placed near the trees, to be used as soon as the sun had warmed the earth. The mound of dirt at the base of each tree was of course levelled down before this dressing was applied. I never afterward purchased stable or stock-yard manure, though I could often have used it to advantage; for I did not think it safe to purchase this kind of fertilizer for a farm where large numbers of animals are kept. The danger from infection is too great. Large quantities of barnyard manure were furnished yearly out of my own pits, and I supplemented it with a good deal of the commercial variety. I try to turn back to the land each year more than I take from it, but I do not dare to go to a stock-yard for any part of my supply. It was not until I had mentally established a quarantine for my hogs that I realized the danger from those six carloads of manure; and I promised myself then that no such breach of quarantine should again occur.

The cows arrived on St. Patrick's Day. Our herd was then composed of the twenty Holstein heifers (coming three years old), and six of the best of the common cows purchased with the farm. Within forty days the herd was increased by the addition of twenty-three calves. Twenty-five were born, but two were dead. Of this number, eighteen were Holsteins eligible for registration, ten heifers, and eight bulls. Each calf was taken from its mother on the third day and fed warm skim-milk from a patent feeder three times a day, all it would drink. When three weeks old, seven of the Holstein calves and the five from the common cows were sent to market. They brought $5.25 each above the expense of selling, or $63 for the bunch. The ten Holstein heifer calves were of course held; and one bull calf, which had a double cross of Pieterje 2d and Pauline Paul, and which seemed an unusually fair specimen, was kept for further development.

The cow barn was finished about April 1st, and shortly after that the herd was established in permanent quarters. As the dairy-house was unfinished, and there was no convenient way of disposing of the milk which now flowed in abundance, I bought a separator (for $200) and sent the cream to a factory, using the fresh skim-milk for the calves and young pigs and chickens.

From March 22, when I began to sell, until May 10, when my dairy-house was in working order, I received $203 for cream. Thompson had sold milk from the old cows, from August to December, 1895, to the amount of $132. This item should have been entered on the credit side for the last year, but as it was not, we will make a note of it here. These are the only sales of milk and cream made from Four Oaks since I bought the land.

The milk supply from my herd started out at a tremendous rate, considering the age of the cows. It must be borne in mind that none of the thoroughbreds was within three years of her (probable) best; yet they were doing nobly, one going as high as fifty-two pounds of milk in one day, and none falling below thirty-six as a maximum. The common cows did nearly as well at first, four of them giving a maximum of thirty-two pounds each in twenty-four hours. It was easy to see the difference between the two sorts, however. The old ones had reached maturity and were doing the best they could; the others were just beginning to manufacture milk, and were building and regulating their machinery for that purpose. The Holsteins, though young, were much larger than the old cows, and were enormous feeders. A third or a half more food passed their great, coarse mouths than their less aristocratic neighbors could be coaxed to eat. Food, of course, is the one thing that will make milk; other things being equal, then the cow that consumes the most food will produce the most milk. This is the secret of the Holsteins' wonderful capacity for assimilating enormous quantities of food without retaining it under their hides in the shape of fat. They have been bred for centuries with the milk product in view, and they have become notable machines for that purpose. They are not the cows for people to keep who have to buy feed in a high market, for they are not easy keepers in any sense; but for the farmer who raises a lot of grain and roughage which should be fed at his own door, they are ideal. They will eat much and return much.

As to feeding for milk, I have followed nearly the same plan through my whole experiment. I keep an abundance of roughage, usually shredded corn, before the cows all the time. When it has been picked over moderately well, it is thrown out for bedding, and fresh fodder is put in its place. The finer forages, timothy, red-top, clover, alfalfa, and oat straw, are always cut fine, wetted, and mixed with grain before feeding. This food is given three times a day in such quantities as will be eaten in forty-five minutes. Green forage takes the place of dry in season, and fresh vegetables are served three times a week in winter. The grain ration is about as follows: By weight, corn and cob meal, three parts; oatmeal, three parts; bran, three parts; gluten meal, two parts; linseed meal, one part. The cash outlay for a ton of this mixture is about $12; this price, of course, does not include corn and oats, furnished by the farm. A Holstein cow can digest fifteen pounds of this grain a day. This means about two and a half tons a year, with a cash outlay of $30 per annum for each head. Fresh water is always given four times a day, and much of the time the cows have ready access to it. In cold weather the water is warmed to about 65° F. The cows are let out in a twenty-acre field for exercise every day, except in case of severe storms. They are fed forage in the open when the weather is fine and insects are not troublesome, and they sometimes sleep in the open on hot nights; but by far the largest part of their time is spent in their own stalls away from chilling winds and biting flies. In their stables they are treated much as fine horses are,—well bedded, well groomed, and well cared for in all ways.

A quiet, darkened stable conduces rumination. Loud talking, shouting, or laughing are not looked upon with favor in our cow barn. On the other hand, continuous sounds, if at all melodious, seem to soothe the animals and increase the milk flow. Judson, who has proved to be our best herdsman, has a low croon in his mouth all the time. It can hardly be called a tune, though I believe he has faith in it, but it has a fetching way with the herd. I have never known him to be quick, sharp, or loud with the cows. When things go wrong, the crooning ceases. When it is resumed, all is well in the cow world. The other man, French, who is an excellent milker, and who stands well with the cows, has a half hiss, half whistle, such as English stable-boys use, except that it runs up and down five notes and is lost at each end.

The cows like it and seem to admire French for his accomplishment even more than Judson, for they follow his movements with evident pleasure expressed in their great ox eyes.

Rigid rules of cleanliness are carried out in every detail with the greatest exactness. The house and the animals are cared for all the time as if on inspection. Before milking, the udders are carefully brushed and washed, and the milker covers himself entirely with a clean apron. As each cow is milked, the milker hangs the pail on a spring balance and registers the exact weight on a blackboard. He then carries the milk through the door that leads to the dairy-house, and pours it into a tank on wheels. This ends his responsibility. The dairymaid is then in charge.

CHAPTER XXV

The Dairymaid

Of course I had trouble in getting a dairymaid. I was not looking for the bouncing, buxom, red-cheeked, arms-akimbo, butter-colored-hair sort. I didn't care whether she were red-cheeked and bouncing or not, but for obvious reasons I didn't want her hair to be butter-colored. What I did want was a woman who understood creamery processes, and who could and would make the very giltest of gilt-edged butter.

I commenced looking for my paragon in January. I interviewed applicants of both sexes and all nationalities, but there was none perfect; no, not one. I was not exactly discouraged, but I certainly began to grow anxious as the time approached when I should need my dairymaid, and need her badly. One day, while looking over the *Rural New Yorker* (I was weaned on that paper), I saw the following advertisement. "Wanted: Employment on a dairy-farm by a married couple who understand the business." If this were true, these two persons were just what I needed; but, was it true? I had tried a score of greater promise and had not found one that would do. Was I to flush two at once, and would they fall to my gun?

A small town in one of the Middle Western states was given as the address, and I wrote at once. My letter was strong in requirements, and asked for particulars as to experience, age, references, and nationality. The reply came promptly, and was more to my liking than any I had received before. Name, French; Americans, newly married, twenty-eight and twenty-six respectively; experience four and three years in creamery and dairy work; references, good; the couple wished to work together to save money to start a dairy of their own. I was pleased with the letter, which was an unusual one to come from native-born Americans. Our people do not often hunt in couples after this manner. I telegraphed them to come to the city at once.

It was late in April when I first saw the Frenches. The man was tall and raw-boned, but good-looking, with a frank manner that inspired confidence. He was a farmer's son with a fair education, who had saved a little money, and had married his wife out of hand lest some one else should carry her off while he was building the nest for her.

"I took her when I could get her," he said, "and would have done it with a two-dollar bill in my pocket rather than have taken chances."

The woman was worthy of such an extreme measure, for she looked capable of caring for both. She was a fine pattern of a country girl, with a head full of good sense, and very useful-looking hands and arms. Her face was good to look upon; it showed strength of character and a definite object in life. She said she understood the creamery processes in all their niceties, and that she could make butter good enough for Queen Victoria.

The proposition offered by this young couple was by far the best I had received, and I closed with them at once. I agreed to pay each $25 a month to start with, and explained my plan of an increasing wage of $1 a month for each period of six months' service. They thought they ought to have $30 level. I thought so, too, if they were as good as they promised. But I had a fondness for my increasing scale, and I held to it. These people were skilled laborers, and were worth more to begin with than ordinary farm hands. That is why I gave them $25 a month from the start. Six hundred dollars a year for a man and wife, with no expense except for clothing, is good pay. They can easily put away $400 out of it, and it doesn't take long to get fore-handed. I think the Frenches have invested $500 a year, on an average, since they came to Four Oaks.

It is now time to get at the dairy-house, since the dairy and the dairymaid are both in evidence. The house was to be on the building line, and both Polly and I thought it should have attractive features. We decided to make it of dark red paving brick. It was to be eighteen feet by thirty, with two rooms on the ground. The first, or south room, ten feet by eighteen, was fitted for storing fruit, and afforded a stairway to the rooms above, which were four in number besides the bath. The larger room was of course the butter factory, and was equipped with up-to-date appliances,—aërator, Pasteurizer, cooler, separator, Babcock tester, swing churn, butter-worker, and so on. The house was to have steep gables and projecting eaves, with a window in each gable, and two dormer windows in each roof. The walls were to be plastered, and the ground floor was to be cement. It cost $1375.

As motive power for the churn and separator, a two-sheep-power treadmill has proved entirely satisfactory. It is worked by two sturdy wethers who are harbored in a pleasant house and run, close to the power-house, and who pay for their food by the sweat of their brows and the wool from their backs. They do not appear to dislike the "demnition grind," which lasts but an hour twice a day; they go without reluctance to the tramp that leads nowhere, and the futile journey which would seem foolish to anything wiser than a sheep. This sheep-power is one of the curios of the place. My grand-girls never lose their interest in it, and it has been photographed and sketched more times than there are fingers and toes on the sheep.

The expenditure for equipment, from separator to sheep, was $354. I made an arrangement with a fancy grocer in the city to furnish him thirty pounds, more or less, of fresh (unsalted) butter, six days in the week, at thirty-three cents a pound, I to pay express charges. I bought six butter-carriers with ice compartments for $3.75 each, $23 in all, and arranged with the express company to deliver my packages to the grocer for thirty cents each. The butter netted me thirty-two cents a pound that year, or about $60 a week.

In July I bought four thoroughbred Holsteins, four years old, in fresh milk, and in October, six more, at an average price of $120 a head,—$1200 in all. These reënforcements made it possible for me to keep my contract with the middleman, and often to exceed it.

The dairy industry was now fairly launched and in working order. It had cost, not to be exact, $7000, and it was reasonably sure to bring back to the farm about $60 a week in cash, besides furnishing butter for the family and an immense amount of skim-milk and butter-milk to feed to the young animals on the place.

CHAPTER XXVI

Little Pigs

By April 1st all my sows had farrowed. There was much variation in the number of pigs in these nineteen litters. One noble mother gave me thirteen, two of which promptly died. Three others farrowed eleven each, and so down to one ungrateful mother who contributed but five to the industry at Four Oaks. The average, however, was good; 154 pigs on April 10th were all that a halfway reasonable factory man could expect.

These youngsters were left with their mothers until eight weeks old; then they were put, in bunches of thirty, into the real hog-house, which was by that time completed. It was 200 feet long and 50 feet wide, with a 10-foot passageway through the length of it. On either side were 10 pens 20 feet by 20, each connected with a run 20 feet by 120. The house stood on a platform or bed of cement 90 by 200 feet, which formed the floor of the house and extended 20 feet outside of each wall, to secure cleanliness and a dry feeding-place in the open. The cement floor was expensive ($1120 as first cost), but I think it has paid for itself several times over in health and comfort to the herd. The structure on this floor was of the simplest; a double wall only five feet high at the sides, shingled roof, broken at the ridge to admit windows, and strong partitions. It cost $3100. As in the brood-sow house, there is a kitchen at the west end. The 150 little pigs made but a small showing in this great house, which was intended to shelter six hundred of all sizes, from the eight-weeks-old baby pig to the nine-months-old three-hundred-pounder ready for market.

Pigs destined for market never leave this house until ripe for killing. At six or seven months a few are chosen to remain on the farm and keep up its traditions; but the great number live their ephemeral lives of eight months luxuriously, even opulently, until they have made the ham and bacon which, poor things, they cannot save, and then pass into the pork barrel or the smoke-house without a sigh of regret. They toil not, neither do they spin; but they have a place in the world's economy, and they fit it perfectly. So long as one animal must eat another, the man animal should thank the hog animal for his generosity.

Now that my big hog-house seemed so empty, I would gladly have sent into the highways and byways to buy young stock to fill it; but I dared not break my quarantine. I could easily have picked up one hundred or even two hundred new-weaned pigs, within six or eight miles of my place, at about $1.50 each, and they would have grown into fat profit by fall; but I would not take a risk that might bear ill fruit. I had slight depressions of spirits when I visited my pig-gery during that summer; but I chirked up a little in the fall, when the brood sows again made good. But more of that anon.

CHAPTER XXVII

Work on the Home Forty

April and May made amends for the rudeness of March, and the ploughs were early afield. Thompson, Zeb, Johnson, and sometimes Anderson, followed the furrows, first in 10 and 11, and lastly in 13. Number 9 had a fair clover sod, and was not disturbed. We ploughed in all about 114 acres, but we did not subsoil. We spent twenty days ploughing and as many more in fitting the ground for seed. The weather was unusually warm for the season, and there was plenty of rain. By the middle of May, oats were showing green in Nos. 8, 10, 11, 12, and 13,—sixty-two acres. The corn was well planted in 15 and the west three-quarters of 14,—eighty-two acres. The other ten acres in the young orchard was planted to fodder corn, sown in drills so that it could be cultivated in one direction.

The ten-acre orchard on the south side of the home lot was used for potatoes, sugar beets, cabbages, turnips, etc., to furnish a winter supply of vegetables for the stock.

The outlook for alfalfa was not bright. In the early spring we fertilized it again, using five hundred pounds to the acre, though it seemed like a conspicuous waste. The warm rains and days of April and May brought a fine crop of weeds; and about the middle of May I turned Anderson loose in the fields with a scythe, and he mowed down everything in sight.

After that things soon began to look better in the alfalfa fields. As the season was favorable, we were able to cut a crop of over a ton to the acre early in July, and nearly as much in the latter part of August. We cut forty tons from these twenty acres within a year from seeding, but I suspect that was unusual luck. I had used thirteen hundred pounds of commercial fertilizer to the acre, and the season was very favorable for the growth of the plant. I have since cut these fields three times each year, with an average yield of five tons to the acre for the whole crop.

I like alfalfa, both as green and as dry forage. When we use it green, we let it lie in swath for twenty-four hours, that it may wilt thoroughly before feeding. It is then fit food for hens, hogs, and, in limited quantities, for cows, and is much relished. When used dry, it is always cut fine and mixed with ground grains. In this shape it is fed liberally to hens and hogs, and also to milch cows; for the latter it forms half of the cut-food ration.

While the crops are growing, we will find time to note the changes on the home lot. Nearly in front of the farm-house, and fifty yards distant, was a space well fitted for the kitchen garden. We marked off a plat two hundred feet by three hundred, about one and a half acres, carted a lot of manure on it, and ploughed it as deep as the subsoiler would reach. This was done as soon as the frost permitted. We expected this garden to supply vegetables and small fruits for the whole colony at Four Oaks. An acre and a half can be made exceedingly productive if properly managed.

Along the sides of this garden we planted two rows of currant and gooseberry bushes, six feet between rows, and the plants four feet apart in the rows. The ends of the plat were left open for convenience in horse cultivation. Ten feet outside these rows of bush fruit was planted a line of quince trees, thirty on each side, and twenty feet beyond these a row of cherry trees, twenty in each row.

Near the west boundary of the home lot, and north of the lane that enters it, I planted two acres of dwarf pear trees—Bartlett and Duchess,—three hundred trees to the acre. I also planted six hundred plum trees—Abundance, Wickson, and Gold—in the chicken runs on lot 4. After May 1, when he was relieved from his farm duties, Johnson had charge of the planting and also of the gardening, and he took up his special work with energy and pleasure.

The drives on the home lot were slightly rounded with ploughs and scraper, and then covered with gravel. The open slope intended for the lawn was now to be treated. It comprised about ten acres, irregular in form and surface, and would require a good deal of work to whip it into shape. A lawn need not be perfectly graded,—in fact, natural inequalities with dips and rises are much more attractive; but we had to take out the asperities. We ploughed it thoroughly, removed all stumps and stones, levelled and sloped it as much as pleased Polly, harrowed it twice a week until late August, sowed it heavily to grass seed, rolled it, and left it.

Polly had the house in her mind's eye. She held repeated conversations with Nelson, and was as full of plans and secrets as she could hold. By agreement, she was to have a free hand to the extent of $15,000 for the house and the carriage barn. I never really examined the plans, though I saw the blue prints of what appeared to be a large house with a driving entrance on the east and a great wide porch along the whole south side. I did not know until it was nearly finished how large, convenient, and comfortable it was to be. A hall, a great living-room, the dining room, a small reception room, and an office, bedroom, and bath for me, were all on the ground floor, besides a huge wing for the kitchen and other useful offices.

Above stairs there was room for the family and a goodly number of friends. We had agreed that the house should be simple in all ways, with no hard wood except floors, and no ornamentation except paint and paper. It must be larger than our needs, for we looked forward to delightful visits from many friends. We were to have more leisure than ever before for social life, and we desired to make the most of our opportunities.

A country house is by all odds the finest place to entertain friends and to be entertained by them. They come on invitation, not as a matter of form, and they stay long enough to put by questions of weather, clothes, and servant-girls, and to get right down to good old-fashioned visiting. Real heart-to-heart talks are everyday occurrences in country visits, while they are exceptional in city calls. We meant to make much of our friends at Four Oaks, and to have them make much of us. We have discovered new values even in old friends, since we began to live with them, weeks at a time, under the same roof. Their interests are ours, and our plans are warmly taken up by them. There is nothing like it among the turmoils and interruptions of town life, and the older we grow the more we need this sort of rest among our friends. The guest book at the farm will show very few weeks, in the past six years, when friends haven't been with us, and Polly and I feel that the pleasure we have received from this source ought to be placed on the credit side of the farm ledger.

Another reason for a company house was that Jack and Jane would shortly be out of school. It was not at all in accord with our plan that they should miss any pleasure by our change. Indeed, we hoped that the change would be to their liking and to their advantage.

CHAPTER XXVIII

Discounting the Market

We broke ground for the house late in May, and Nelson said that we should be in it by Thanksgiving Day. Soon after the plans were settled Polly informed me that she should not spend much money on the stable.

"Can't do it," she said, "and do what I ought to on the house. I will give you room for six horses; the rest, if you have more, must go to the farm barn. I cannot spend more than $1100 or $1200 on the barn."

Polly was boss of this department, and I was content to let her have her way. She had already mulcted me to the extent of $436 for trees, plants, and shrubs which were even then grouped on the lawn after a fashion that pleased her. I need not go into the details of the lawn planting, the flower garden, the pergola, and so forth. I have a suspicion that Polly has in mind a full account of the "fight for the home forty," in a form greatly better than I could give it, and it is only fair that she should tell her own story. I am not the only one who admires her landscape, her flower gardens, and her woodcraft. Many others do honor to her tastes and to the evidence of thought which the home lot shows. She disclaims great credit, for she says, "One has only to live with a place to find out what it needs."

As I look back to the beginning of my experiment, I see only one bit of good luck that attended it. Building material was cheap during the months in which I had to build so much. Nothing else specially favored me, while in one respect my experiment was poorly timed. The price of pork was unusually low. For three years, from 1896, the price of hogs never reached $5 per hundred pounds in our market,—a thing unprecedented for thirty years. I never sold below three and a half cents, but the showing would have been wonderfully bettered could I have added another cent or two per pound for all the pork I fattened. The average price for the past twenty-five years is well above five cents a pound for choice lots. Corn and all other foods were also cheap; but this made little difference with me, because I was not a seller of grain.

In 1896 I was, however, a buyer of both corn and oats. In September of that year corn sold on 'Change at 19-1/2 cents a bushel, and oats at 14-3/4. These prices were so much below the food value of these grains that I was tempted to buy. I sent a cash order to a commission house for five thousand bushels of each. I stored this grain in my granary, against the time of need, at a total expense of $1850,—21 cents a bushel for corn and 16 for oats. I had storage room and to spare, and I knew that I could get more than a third of a cent out of each pound of corn, and more than half a cent out of each pound of oats. I recalled the story of a man named Joseph who did some corn business in Egypt a good many years ago, much in this line, and who did well in the transaction. There was no dream of fat kine in my case; but I knew something of the values of grains, and it did not take a reader of riddles to show me that when I could buy cheaper than I could raise, it was a good time to purchase.

As I said once before, there have been no serious crop failures at Four Oaks,—indeed, we can show better than an average yield each year; but this extra corn in my cribs has given me confidence in following my plan of very liberal feeding. With this grain on hand I was able to cut twenty acres of oats in Nos. 10 and 11 for forage. This was done when the grain was in the milk, and I secured about sixty tons of excellent hay, much loved by horses. We got from No. 9 a little less than twelve tons of clover,—alfalfa furnished forty tons; and there was nearly twenty tons of old hay left over from that originally purchased. With all this forage, good of its kind, there was, however, no timothy or red top, which is by all odds the best hay for horses. I determined to remedy this lack before another year. As soon as the oats were off lots 10 and 11, they were ploughed and crossed with the disk harrow. From then until September 1, these fields were harrowed each week in half lap, so that by the time we were ready to seed them they were in excellent condition and free from weeds. About September 1 they were sown to timothy and red top, fifteen pounds each to the acre, top-dressed with five hundred pounds of fertilizer, harrowed once more, rolled, and left until spring, when another dose of fertilizer was used.

I wished to establish twenty acres of timothy and as much alfalfa, to furnish the hay supply for the farm. With one hundred tons of alfalfa and sixty of timothy, which I could reasonably expect, I could get on splendidly.

From the first I have practised feeding my hay crop for immediate returns. The land receives five hundred pounds of fertilizer per acre when it is sown, a like amount again in the spring, and, as soon as a crop is cut, three hundred pounds an acre more. This usually gives a second crop of timothy about September 1, if the season is at all favorable. The alfalfa is cut at least three times, and for each cutting it receives three hundred pounds of plant food per acre. In the course of a year I spend from $10 to $12 an acre for my grass land. In return I get from each acre of timothy, in two cuttings, about three and a half tons; worth, at an average selling price, $12 a ton. The alfalfa yields nearly five tons per acre, and has a feeding value of $10 a ton. I have sold timothy hay a few times, but I feel half ashamed to say so, for it is against my view of justice to the land. I find oat hay cheaper to raise than timothy, and, as it is quite as well liked by the horses, I have been tempted to turn a part of my timothy crop into money directly from the field.

CHAPTER XXIX

From City to Country

In early July I went through my young orchard, which had been cut back so ruthlessly the previous autumn, and carefully planned a head for each tree. Quite a bunch of sprouts had started from near the top of each stub, and were growing luxuriantly. Out of each bunch I selected three or four to form the head; the rest were rubbed off or cut out with a sharp knife or pruning shears. It surprised me to see what a growth some of these sprouts had made; sixteen or eighteen inches was not uncommon. Big roots and big bodies were pushing great quantities of sap toward the tops.

Of course I bought farm machinery during this first season,— mower, reaper, corn reaper, shredder, and so on. In October I took account of expenditures for machinery, grass seed, and fertilizer, and found that I had invested $833. I had also, at an expense of $850, built a large shed or tool-house for farm implements. It is one of the rules at Four Oaks to grease and house all tools when not in actual use. I believe the observation of this rule has paid for the shed.

In October 1896 I had a good offer for my town house, and accepted it. I had purchased the property eleven years before for $22,000, but, as it was in bad condition, I had at once spent $9000 on it and the stable. I sold it for $34,000, with the understanding that I could occupy it for the balance of the year if I wished.

After selling the house, I calculated the cost of the elementary necessities, food and shelter, which I had been willing to pay during many years of residence in the city. The record ran about like this:—

Interest at 5% on house valued at $34,000	$1700.00
Yearly taxes on same	340.00
Insurance	80.00
Fuel and light	250.00
Wages for one man and three women	1200.00
Street sprinkling, watchman, etc.	90.00
Food, including water, ice, etc.	1550.00
Making a total of	$5210.00

It cost me $100 a week to shelter and feed my family in the city. This, of course, took no account of personal expenses,—travel, sight-seeing, clothing, books, gifts, or the thousand and one things which enter more or less prominently into the everyday life of the family.

If the farm was to furnish food and shelter for us in the future, it would be no more than fair to credit it with some portion of this expenditure, which was to cease when we left the city home. What portion of it could be justly credited to the farm was to be decided by comparative comforts after a year of experience. I did not plan our exodus for the sake of economy, or because I found it necessary to retrench; our rate of living was no higher than we were willing and able to afford. Our object was to change occupation and mode of life without financial loss, and without moulting a single comfort. We wished to end our days close to the land, and we hoped to prove that this could be done with both grace and profit. I had no desire to lose touch with the city, and there was no necessity for doing so. Four Oaks is less than an hour from the heart of town. I could leave it, spend two or three hours in town, and be back in time for luncheon without special effort; and Polly would think nothing of a shopping trip and friends home with her to dinner. The people of Exeter were nearly all city people who were so fortunate as not to be slaves to long hours. They were rich by work or by inheritance, and they gracefully accepted the *otium cum dignitate* which this condition permitted. Social life was at its best in Exeter, and many of its people were old acquaintances of ours. A noted country club spread its broad acres within two miles of our door, and I had been favorably posted for membership. It did not look as though we should be thrust entirely upon our own resources in the country; but at the worst we had resources within our own walls and fences that would fend off all but the most violent attacks of ennui.

We were both keenly interested in the experiment. Nothing that happened on the farm went unchallenged. The milk product for the day was a thing of interest; the egg count could not go unnoted; a hatch of chickens must be seen before they left the incubator; a litter of new-born pigs must be admired; horses and cows were forever doing things which they should or should not do; men and maids had griefs and joys to share with mistress or Headman; flowers were blooming, trees were leafing, a robin had built in the black oak, a gopher was tunnelling the rose bed,—a thousand things, full of interest, were happening every day. As a place where things the most unexpected do happen, recommend me to a quiet farm.

But we were not to depend entirely upon outside things for diversion. Books we had galore, and we both loved them. Many a charming evening have I spent, sometimes alone, more often with two or three congenial friends, listening to Polly's reading. This is one of her most delightful accomplishments. Her friends never tire of her voice, and her voice never tires of her friends. We all grow lazy when she is about; but there are worse things than indolence. No, we did not mean to drop out of anything worth while; but we were pretty well provisioned against a siege, if inclement weather or some other accident should lock us up at the farm.

To keep still better hold of the city, I suggested to Tom and Kate that they should keep open house for us, or any part of us, whenever we were inclined to take advantage of their hospitality. This would give us city refuge after late functions of all sorts. The plan has worked admirably. I devote $1200 a year out of the $5200 of food-and-shelter money to the support of our city shelter at Kate's house, and the balance, $4000, is entered at the end of each year on the credit side of the farm ledger. Nor do I think this in any way unjust. We do not expect to get things for nothing, and we do not wish to. If the things we pay for now are as valuable as those we paid for six or eight years ago, we ought not to find fault with an equal price. I have repeatedly polled the family on this question, and we all agree that we have lost nothing by the change, and that we have gained a great deal in several ways. Our friends are of like opinion; and I am therefore justified in crediting Four Oaks with a considerable sum for food and shelter. We have bettered our condition without foregoing anything, and without increasing our expenses. That is enough.

CHAPTER XXX

Autumn Reckoning

We harvested the crops in the autumn of 1896, and were thankful for the bountiful yield. Nearly sixteen hundred bushels of oats and twenty-seven hundred bushels of corn made a proud showing in the granary, when added to its previous stock. The corn fodder, shredded by our own men and machine, made the great forage barn look like an overflowing cornucopia, and the only extra expense attending the harvest was $31 paid for threshing the oats.

Three important items of food are consumed on the farm that have to be purchased each year, and as there is not much fluctuation in the price paid, we may as well settle the per capita rate for the milch cows and hogs for once and all. At each year's end we can then easily find the cash outlay for the herds by multiplying the number of stock by the cost of keeping one.

My Holstein cows consume a trifle less than three tons of grain each per year,—about fifteen pounds a day. Taking the ration for four cows as a matter of convenience, we have: corn and cob meal, three tons, and oatmeal, three tons, both kinds raised and ground on the farm, and not charged in this account; wheat bran, three tons at $18, $54; gluten meal, two tons at $24, $48; oil meal, one ton, $26; total cash outlay for four cows, $128, or $32 per head. This estimate is, however, about $2 too liberal. We will, hereafter, charge each milch cow $30, and will also charge each hog fattened on the place $1 for shorts and middlings consumed. This is not exact, but it is near enough, and it greatly simplifies accounts.

As I kept twenty-six cows ten months, and ten more for an average of four and a half months, the feeding for 1896 would be equivalent to one year for thirty cows, or $900. To this add $120 for swine food and $25 for grits and oyster shells for the chickens, and we have $1045 paid for food for stock. Shoeing the horses for the year and repairs to machinery cost $157. The purchased food for eight employees for twelve months and for two additional ones for eight months, amounted to $734. The wage account, including $50 extra to Thompson, was $2358.

A second hen-house, a duplicate of the first, was built before October. It was intended that each house should accommodate four hundred laying hens. We have now on the place five of these houses; but only two of them, besides the incubator and the brooder-house, were built in 1896. As offset to the heavy expenditure of this year, I had not much to show. Seven hundred cockerels were sold in November for $342. In October the pullets began laying in desultory fashion, and by November they had settled down to business; and that quarter they gave me 703 dozen eggs to sell. As these eggs were marketed within twenty-four hours, and under a guarantee, I had no difficulty in getting thirty cents a dozen, net. November eggs brought $211, and the December out-put, $252. I sold 600 bushels of potatoes for $150, and the apples from 150 of the old trees (which, by the way, were greatly improved this year) brought $450 on the trees.

The cows did well. In the thirty-three weeks from May 12 to December 31, I sold a little more than 6600 pounds of butter, which netted me $2127.

We had 122 young hogs to sell in December. They had been crowded as fast as possible to make good weight, and they went to market at an average of 290 pounds a head. The price was low, but I got the top of the market,—$3.55 a hundred, which amounted to $1170 after paying charges. I had reserved twenty-five of the most likely young sows to stay on the farm, and had transferred eight to the village butcher, who was to return them in the shape of two barrels of salt pork, thirty-two smoked hams and shoulders, and a lot of bacon.

The old sows farrowed again in September and early October, and we went into the winter with 162 young pigs. I get these details out of the way now in order to turn to the family and the social side of life at Four Oaks.

CHAPTER XXXI

The Children

The house did not progress as fast as Nelson had promised, and it was likely to be well toward Christmas before we could occupy it. As the days shortened, Polly and I found them crowded with interests. Life at Four Oaks was to mean such a radical change that we could not help speculating about its influence upon us and upon the children. Would it be satisfactory to us and to them? Or should we find after a year or two of experiment that we had been mistaken in believing that we could live happier lives in the country than in town? A year and a half of outdoor life and freedom from professional responsibilities had wrought a great change in me. I could now eat and sleep like a hired man, and it seemed preposterous to claim that I was going to the country for my health. My medical adviser, however, insisted that I had not gotten far enough away from the cause of my breakdown, and that it would be unwise for me to take up work again for at least another year. In my own mind there was a fixed opinion that I should never take it up again. I loved it dearly; but I had given long, hard service to it, and felt that I had earned the right to freedom from its exacting demands. I have never lost interest in this, the noblest of professions, but I had done my share, and was now willing to watch the work of others. In my mind there was no doubt about the desirability of the change. I have always loved the thought of country life, and now that my thoughts were taking material shape, I was keen to push on. Polly looked toward the untrammelled life we hoped to lead with as great pleasure as I.

But how about the children? Would it appeal to them with the same force as to us? The children have thus far been kept in the background. I wanted to start my factory farm and to get through with most of its dull details before introducing them to the reader, lest I should be diverted from the business to the domestic, or social, proposition.

The farm is laid by for the winter, and most of the details needed for a just comprehension of our experiment have been given. From this time on we will deal chiefly with results. We will watch the output from the factory, and commend or find fault as the case may deserve.

The social side of life is quite as important as the commercial, for though we gain money, if we lose happiness, what profit have we? Let us study the children to see what chances for happiness and good fellowship lie in them.

Kate is our first-born. She is a bright, beautiful woman of five-and-twenty, who has had a husband these six years, one daughter for four years, and, wonderful to relate, another daughter for two years. She is quick and practical, with strong opinions of her own, prompt with advice and just as prompt with aid; a woman with a temper, but a friend to tie to in time of stress. She has the education of a good school, and what is infinitely better, the cultivation of an observing mind. She is quick with tongue and pen, but her quickness is so tempered by unquestioned friendliness that it fastens people to her as with a cord. She overflows with interests of every description, but she is never too busy to listen sympathetically to a child or a friend. She is the practical member of the family, and we rarely do much out of the ordinary without first talking it over with Kate.

Tom Hamilton, her husband, is a young man who is getting on in the world. He is clever in his profession, and sure to succeed beyond the success of most men. He is quiet in manner, but he seems to have a way of managing his quick, handsome wife, which is something of a surprise to me, and to her also, I fancy. They are congenial and happy, and their children are beings to adore. Tom and Kate are to live in town. They are too young for the joys of country life, and must needs drag on as they are, loved and admired by a host of friends. They can, and will, however, spend much time at Four Oaks; and I need not say they approved our plans.

Jack is our second. He was a junior at Yale, and I am shy of saying much about him lest I be accused of partiality. Enough to say that he is tall, blond, handsome, and that he has gentle, winning ways that draw the love of men and women. He is a dreamer of dreams, but he has a sturdy drop of Puritan blood in his veins that makes him strong in conviction and brave in action. Jack has never caused me an hour of anxiety, and I was ever proud to see him in any company.

Concerning Jane, I must be pardoned in advance for a father's favoritism. She is my youngest, and to me she seems all that a father could wish. Of fair height and well moulded, her physique is perfect. Good health and a happy life had set the stamp of superb womanhood upon her eighteen years. Any effort to describe her would be vain and unsatisfactory. Suffice it to say that she is a pure blonde, with eyes, hair, and skin just to my liking. She is quiet and shy in manner, deliberate in speech, sensitive beyond measure, wise in intuitive judgment, clever in history and literature, but always a little in doubt as to the result of putting seven and eight together, and not unreasonably dominated by the rules of orthography. She is fond of outdoor life, in love with horses and dogs, and withal very much of a home girl. Every one makes much of Jane, and she is not spoiled, but rather improved by it. She was in her second year at Farmington, and, like all Farmington students, she cared more for girls than for boys.

These were the children whom I was to transport from the city, where they were born, to the quiet life at Four Oaks. After carefully taking their measures, I felt little hesitation about making the change. They, of course, had known of the plan, and had often been to the farm; but they were still to find out what it really meant to live there. A saddle horse and dogs galore would square me with Jane, beyond question; but what about Jack? Time must decide that. His plan of life was not yet formed, and we could afford to wait. We did not have much time in which to weigh these matters, for the Christmas holidays were near, and the youngsters would soon be home. We planned to be settled in the new house when they arrived.

CHAPTER XXXII

The Home-Coming

In arranging to move my establishment I was in a quandary as to what it was best to do for a coachman. Lars had been with me fifteen years. He came a green Swedish lad, developed into a first-class coachman, married a nice girl—and for twelve years he and his wife lived happily in the rooms above my stable. Two boys were born to them, and these lads were now ten and twelve years of age. Shortly after I bought the farm Lars was so unfortunate as to lose his good wife, and he and the boys were left forlorn. A relative came and gave them such care as she could, but the mother and wife was missed beyond remedy. In his depression Lars took to drink, and things began to go wrong in the stable. He was not often drunk, but he was much of the time under the influence of alcohol, and consequently not reliable. I had done my best for the poor fellow, and he took my lectures and chidings in the way they were intended, and, indeed, he tried hard to break loose from the one bad habit, but with no good results. His evil friends had such strong hold on him that they could and would lead him astray whenever there was opportunity. Polly and I had many talks about this matter. She was growing timid under his driving, and yet she was attached to him for long and faithful service.

"Let's chance it," she said. "If we get him away from these people who lead him astray, he may brace up and become a man again."

"But what about the boys, Polly?" said I.

"We ought to be able to find something for the boys to do on the farm, and they can go to school at Exeter. Can't they drive the butter-cart out each morning and home after school? They're smart chaps, you know, and used to doing things."

Polly had found a way, and I was heartily glad of it, for I did not feel like giving up my hold on the man and the boys. Lars was glad of the chance to make good again, and he willingly agreed to go. He was to receive $23 a month. This was less than he was getting in the city, but it was the wage which we were paying that year at the farm, and he was content; for the boys were each to receive $5 a month, and to be sent to school eight months a year for three years.

This matter arranged, we began to plan for the moving. I had five horses in my stable,—a span of blacks for the carriage and three single drivers. Besides the horses, harness, and equipment, there was a large carriage, a brougham, a Goddard phæton, a runabout, and a cart. I exchanged the brougham and the Goddard for a station wagon and a park phæton, as more suitable for country use.

The barn equipment was all sent in one caravan, Thompson and Zeb coming into town to help Lars drive out. Our lares and penates were sent by freight on December 17. Polly had managed to coax another thousand dollars out of me for things for the house; and these, with the furniture from our old home, made a brave showing when we gathered around the big fire in the living room, December 22, for our first night in the country.

Tom, Kate, and the grand-girls were with us to spend the holidays, and so, too, was the lady whom we call Laura. I shall not try to say much about Laura. She was a somewhat recent friend. How we ever came to know her well, was half a mystery; and how we ever got on before we knew her well, was a whole one.

Roaring fires and shaded lamps gave an air of homelike grace to our new house, and we decided that we would never economize in either wood or oil; they seemed to stir the home spirit more than ever did coal or electricity.

The day had been a busy one for the ladies, but they were pleased with results as they looked around the well-ordered house and saw the work of their hands. Before separating for the night, Kate said:—

"I'm going to town to-morrow, and I'll pick up Jane and Jack in time to take the four o'clock train out. Papa will meet us at the station, and Momee will greet us at the doorstep. Make an illumination, Momee, and we will carry them by storm. Tom will have to take a later train, but he will be here in time for dinner."

The afternoon of the 23d, the children came, and there was no failure in Kate's plan. The youngsters were delighted with everything. Jane said:—

"I always wanted to live on a farm. I can have a saddle horse now, and keep as many dogs as I like, can't I, Dad?"

"You shall have the horse, and the dogs, too, when you come to stay."

"Daddy," said Jack, "this will be great for you. Let me finish at an agricultural college, so that I can be of some practical help."

"Not on your life, my son! What your daddy doesn't know about farming wouldn't spoil a cup of tea! While you are at home I will give you daily instruction in this most wholesome and independent business, which will be of incalculable benefit to you, and which, I am frank to say, you cannot get in any agricultural college. College, indeed! I have spent thousands of hours in dreaming and planning what a farm should be like! Do you suppose I am going to let these visions become contaminated by practical knowledge? Not by a long way! I have, in the silent watches of the night, reduced the art to mathematical exactness, and I can show you the figures. Don't talk to me about colleges!"

After supper we took the children through the house. Every part was inspected, and many were the expressions of pleasure and admiration. They were delighted with their rooms, and apparently with everything else. We finally quieted down in front of the open fire and discussed plans for the holidays. The children decided that it must be a house party.

"Florence Marcy is with an aunt for whom she doesn't particularly care, and Minnie will just jump at the chance of spending a week in the country," said Jane.

"You can invite three girls, and Jack can have three men. Of course Jessie Gordon will be here. We will drive over in the morning and make sure of her."

"Jack, whom will you ask? Get some good men out here, won't you?"

"The best in the world, little sister, and you will have to keep a sharp lookout or you will lose your heart to one of them. Frank Howard will count it a lark. He has stuck to the "business" as faithfully as if he were not heir to it, and he will come sure to-morrow night. Dear old Phil—my many years' chum—will come because I ask him. These two are all right, and we can count on them. The other one is Jim Jarvis,—the finest man in college."

"Tell us about him, Jack."

"Jarvis's father lives in Montana, and has a lot of gold mines and other things to keep him busy. He doesn't have time to pay much attention to his son, who is growing up after his own fashion. Jim's mother is dead, and he has neither brother nor sister,—nothing but money and beauty and health and strength and courage and sense and the stanchest heart that ever lifted waistcoat! He has been on the eleven three years. They want him in the boat, but he'll not have it; says it's not good work for a man. He's in the first division, well toward the front, too, and in the best society. He's taken a fancy to me, and I'm dead gone on him. He's the man for you to shun, little woman, unless you wish to be led captive."

"There are others, Jack, so don't worry about me. But do you think you can secure this paragon?"

"Not a doubt of it! I'll wire him in the morning, and he'll be here as soon as steam can bring him; he's my best chum, you know."

This would make our party complete. We were all happy and pleased, and the evening passed before we knew it.

CHAPTER XXXIII

Christmas Eve

The next day was a busy one for all of us. Polly and Jane drove to the Gordons and secured Miss Jessie, and then Jane went to town to fetch her other friends. Jack went with her, after having telegraphed to Jim Jarvis. They all came home by mid-afternoon, just as a message came from Jarvis: "Will be on deck at six."

Florence Marcy and Minnie Henderson were former neighbors and schoolmates of Jane's. They were fine girls to look at and bright girls to talk with; blondes, eighteen, high-headed, full of life, and great girls for a house party. Phil and Frank were good specimens of their kinds. Frank was a little below medium height, slight, blond, vivacious to a degree, full of fun, and the most industrious talker within miles; he would "stir things up" at a funeral. Phil Stone was tall, slender, dark, quiet, well-dressed, a good dancer, and a very agreeable fellow in the corner of the room, where his low musical voice was most effective.

Jessie Gordon came at five o'clock. We were all very fond of Jessie, and who could help it? She was tall (considerably above the average height), slender, straight as an arrow, graceful in repose and in motion. She carried herself like a queen, with a proud kind of shyness that became her well. Her head was small and well set on a slender neck, her hair dark, luxurious, wavy, and growing low over a broad forehead, her eyes soft brown, shaded by heavy brows and lashes. She had a Grecian nose, and her mouth was a shade too wide, but it was guarded by singularly perfect and sensitive lips. Her chin was pronounced enough to give the impression of firmness; indeed, save for the soft eyes and sensitive mouth, firmness predominated. She was not a great talker, yet every one loved to listen to her. She laughed with her eyes and lips, but rarely with her voice. She enjoyed intensely, and could, therefore, suffer intensely. She was a dear girl in every way.

All was now ready for the début of Jack's paragon. Jack had driven to the station to fetch him, and presently the sound of wheels on the gravel drive announced the arrival of the last guest. I went into the hall to meet the men.

"Daddy, I want you to know my chum, Jim Jarvis,—the finest all-round son of old Eli. Jarvis, this is my daddy,—the finest father that ever had son!"

"I'm right glad to meet you, Mr. Jarvis; your renown has preceded you."

"I fear, Doctor, it has *exceeded* me as well. Jack is not to be trusted on all subjects. But, indeed, I thank you for your hospitality; it was a godsend to me."

As we entered the living room, Polly came forward and I presented Jarvis to her.

"You are more than welcome, Mr. Jarvis! Jack's 'best friend' is certain of a warm corner at our fireside."

"Madam, I find no word of thanks, but I *do* thank you. I have envied Jack his home letters and the evidences of mother care more than anything else,—and God knows there are enough other things to envy him for. I have no mother, and my father is too busy to pay much attention to me. I wish you would adopt me; I'll try to rival Jack in all that is dutiful."

She did adopt him then and there, for who could refuse such a son! Brown hair, brown eyes, brown skin, a frank, rugged, clean-shaven face, features strong enough to excite criticism and good enough to bear it; broad-shouldered, deep-chested, strong in arm and limb, he carried his six feet of manhood like an Apollo in tweeds. He was introduced to the girls,—the men he knew,—but he was not so quick in his speeches to them. Our Hercules was only mildly conscious of his merits, and was evidently relieved when Jack hurried him off to his room to dress for dinner. When he was fairly out of hearing there was a chorus of comments. The girls all declaimed him handsome, and the boys said:—

"That isn't the best of it,—he's a *trump*! Wait till you know him."

Jane was too loyal to Jack to admit that his friend was any handsomer or in any way a finer fellow than her brother.

"Who said he was?" said Frank, "Jack Williams is out and out the finest man I know. We were sizing him up by such fellows as Phil and me."

"Jack's the most popular man at Yale," said Phil, "but he's too modest to know it; Jarvis will tell you so. He thinks it's a great snap to have Jack for his chum."

These things were music in my ears, for I was quite willing to agree with the boys, and the mother's eyes were full of joy as she led the way to the dining room. That was a jolly meal. Nothing was said that could be remembered, and yet we all talked a great deal and laughed a great deal more. City, country, farm, college, and seminary were touched with merry jests. Light wit provoked heavy laughter, and every one was the better for it. It was nine o'clock before we left the table. I heard Jarvis say:—

"Miss Jane, I count it very unkind of Jack not to have let me go to Farmington with him last term. He used to talk of his 'little sister' as though she were a miss in short dresses. Jack is a deep and treacherous fellow!"

"Rather say, a very prudent brother," said Jane. "However, you may come to the Elm Tree Inn in the spring term, if Jack will let you."

"I'll work him all winter," was Jarvis's reply.

CHAPTER XXXIV

Christmas

Christmas light was slow in coming. There was a hush in the air as if the earth were padded so that even the footsteps of Nature might not be heard. Out of my window I saw that a great fall of snow had come in the night. The whole landscape was covered by fleecy down—soft and white as it used to be when I first saw it on the hills of New England. No wind had moved it; it lay as it fell, like a white mantle thrown lightly over the world. Great feathery flakes filled the air and gently descended upon the earth, like that beautiful Spirit that made the plains of Judea bright two thousand years ago. It seemed a fitting emblem of that nature which covered the unloveliness of the world by His own beauty, and changed the dark spots of earth to pure white.

It was an ideal Christmas morning,—clean and beautiful. Such a wealth of purity was in the air that all the world was clothed with it. The earth accepted the beneficence of the skies, and the trees bent in thankfulness for their beautiful covering. It was a morning to make one thoughtful,—to make one thankful, too, for home and friends and country, and a future that could be earned, where the white folds of usefulness and purity would cover man's inheritance of selfishness and passion.

For an hour I watched the big flakes fall; and, as I watched, I dreamed the dream of peace for all the world. The brazen trumpet of war was a thing of the past. The white dove of peace had built her nest in the cannon's mouth and stopped its awful roar. The federation of the world was secured by universal intelligence and community of interest. Envy and selfishness and hypocrisy, and evil doing and evil speaking, were deeply covered by the snowy mantle that brought "peace on earth and good will to men."

My dream was not dispelled by any rude awakening. As the house threw off the fetters of the night and gradually struggled into activity, it was in such a fresh and loving manner and with such thoughtful solicitude for each member of our world, that I walked in my dream all day.

The snow fell rapidly till noon, and then the sun came forth from the veil of clouds and cast its southern rays across the white expanse with an effect that drew exclamations of delight from all who had eyes to see. No wind stirred the air, but ever and anon a bright avalanche would slide from bough or bush, sparkle and gleam as the sun caught it, and then sink gently into the deep lap spread below. The bough would spring as if to catch its beautiful load, and, failing in this, would throw up its head and try to look unconcerned,—though quite evidently conscious of its bereavement.

The appearance of the sun brought signs of life and activity. The men improvised a snow-plough, the strong horses floundering in front of it made roads and paths through the two feet of feathers that hid the world.

After lunch, the young people went for a frolic in the snow. Two hours later the shaking of garments and stamping of feet gave evidence of the return of the party. Stepping into the hall I was at once surrounded by the handsomest troupe of Esquimaux that ever invaded the temperate zone. The snow clung lovingly to their wet clothing and would not be shaken off; their cheeks were flushed, their eyes bright, and their voices pitched at an out-of-doors key.

"Away to your rooms, every one of you, and get into dry clothes," said I. "Don't dare show yourselves until the dinner bell rings. I'll send each of you a hot negus,—it's a prescription and must be taken; I'm a tyrant when professional."

We saw nothing more of them until dinner. The young ladies came in white, with their maiden shoulders losing nothing by contact with their snow-white gowns. All but Miss Jessie, whose dress was a pearl velvet, buttoned close to her slender throat. I loved this style best, but I could never believe that anything could be prettier than Jane's white shoulders.

The table was loaded, as Christmas tables should be, and, as I asked God's blessing on it and us, the thought came that the answer had preceded the request and that we were blessed in unusual degree.

After dinner the rugs in the great room were rolled up, and the young folks danced to Laura's music, which could inspire unwilling feet. But there were none such that night. Tom and Kate led off in the newest and most fantastic waltz, others followed, and Polly and I were the only spectators. An hour of this, and then we gathered around the hearth to hear Polly read "The Christmas Carol." No one reads like Polly. Her low, soft voice seems never to know fatigue, but runs on like a musical brook. When the reading was over, a hush of satisfied enjoyment had taken possession of us all. It was not broken when Miss Jessie turned to the piano and sang that glorious hymn, "Lead, Kindly Light." Jack was close beside her, his blue eyes shining with an appreciation of which any woman might be proud, and his baritone in perfect harmony with her rich contralto. The young ladies took the higher part, Frank added his tenor, and even Phil and I leaned heavily on Jarvis's deep bass. My effort was of short duration; a lump gathered in my throat that caused me to turn away. Polly was searching fruitlessly for something to dry the tears that overran her eyes, and I was able to lend her aid, but the accommodation was of the nature of a "call loan."

As we separated for the night, Jarvis said: "Lady mother, this day has been a revelation to me. If I live a hundred years, I shall never forget it." I was slow in bringing it to a close. As I loitered in my room, I heard the shuffling of slippered feet in the hall, and a timid knock at Polly's door. It was quickly opened for Jane and Jessie, and I heard sobbing voices say:—

"Momee, we want to cry on your bed," and, "Oh, Mrs. Williams, why can't all days be like this!"

Polly's voice was low and indistinct, but I know that it carried strong and loving counsel; and, as I turned to my pillow, I was still dreaming the dream of the morning.

CHAPTER XXXV

We Close the Books For '96

The morning after Christmas broke clear, with a wind from the south that promised to make quick work of the snow. The young people were engaged for the evening, as indeed for most evenings, in the hospitable village, and they spent the day on the farm as pleased them best.

There were many things to interest city-bred folk on a place like Four Oaks. Everything was new to them, and they wanted to see the workings of the factory farm in all its detail. They made friends with the men who had charge of the stock, and spent much time in the stables. Polly and I saw them occasionally, but they did not need much attention from us. We have never found it necessary to entertain our friends on the farm. They seem to do that for themselves. We simply live our lives with them, and they live theirs with us. This works well both for the guests and for the hosts.

The great event of the holiday week was a New Year Eve dance at the Country Club. Every member was expected to appear in person or by proxy, as this was the greatest of many functions of the year.

Sunday was warm and sloppy, and little could be done out of doors. Part of the household were for church, and the rest lounged until luncheon; then Polly read "Sonny" until twilight, and Laura played strange music in the half-dark.

The next day the men went into town to look about, and to lunch with some college chums. As they would not return until five, the ladies had the day to themselves. They read a little, slept a little, and talked much, and were glad when five o'clock and the men came. Tea was so hot and fragrant, the house so cosey, and the girls so pretty, that Jack said:—

"What chumps we men were to waste the whole day in town!"

"And what do you expect of men, Mr. Jack?" said Jessie.

"Yes, I know, the old story of pearls and swine, but there are pearls and pearls."

"Do you mean that there are more pearls than swine, Mr. Jack?

For, if you do, I will take issue with you."

"If I am a swine, I will be an æsthetic one and wear the pearl that comes my way," said Jack, looking steadily into the eyes of the high-headed girl.

"Will you have one lump or two?"

"One," said Jack, as he took his cup.

The last day of the year came all too quickly for both young and old at Four Oaks. Polly and I went into hiding in the office in the afternoon to make up the accounts for the year. As Polly had spent the larger lump sum, I could face her with greater boldness than on the previous occasion. Here is an excerpt from the farm ledger:—

Expended in 1896	$43,309
Interest on previous account	2,200
Total	$45,509
Receipts	5,105
Net expense	$40,404
Previous account	44,000
	$84,404

The farm owes me a little more than $84,000. "Not so good as I hoped, and not so bad as I feared," said Polly. "We will win out all right, Mr. Headman, though it does seem a lot of money."

"Like the Irishman's pig," quoth I. "Pat said, 'It didn't weigh nearly as much as I expected, but I never thought it would.'"

There was little to depress us in the past, and nothing in the present, so we joined the young people for the dance at the Club.

CHAPTER XXXVI

Our Friends

After our guests had departed, to college or school or home, the house was left almost deserted. We did not shut it up, however. Fires were bright on all hearths, and lamps were kept burning. We did not mean to lose the cheeriness of the house, though much of the family had departed. For a wonder, the days did not seem lonesome. After the fist break was over, we did not find time to think of our solitude, and as the weeks passed we wondered what new wings had caused them to fly so swiftly. Each day had its interests of work or study or social function. Stormy days and unbroken evenings were given to reading. We consumed many books, both old and new, and we were not forgotten by our friends. The dull days of winter did not drag; indeed, they were accepted with real pleasure. Our lives had hitherto been too much filled with the hurry and bustle inseparable from the fashionable existence-struggle of a large city to permit us to settle down with quiet nerves to the real happiness of home. So much of enjoyment accompanies and depends upon tranquillity of mind, that we are apt to miss half of it in the turmoil of work-strife and social-strife that fill the best years of most men and women.

It is a pity that all overwrought people cannot have a chance to relax their nerves, and to learn the possibilities of happiness that are within them. Most of the jars and bickerings of domestic life, most of the mental and moral obliquities, depend upon threadbare nerves, either inherited or uncovered by friction incident to getting on in the world. I never understood the comforts that follow in the wake of a quiet, unambitious life, until such a life was forced upon me. When you discover these comforts for the first time, you marvel that you have foregone them so long, and are fain to recommend them to all the world.

Polly and I had gotten on reasonably well up to this time; but before we became conscious of any change, we found ourselves drawn closer together by a multitude of small interests common to both. After twenty-five years of married life it will compensate any man to take a little time from business and worry that he may become acquainted with his wife. A few fortunate men do this early in life, and they draw compound interest on the investment; but most of us feel the cares of life so keenly that we take them home with us to show in our faces and to sit at our tables and to blight the growth of that cheerful intercourse which perpetuates love and cements friendship in the home as well as in the world.

There were no serious cares nowadays, and time passed so smoothly at Four Oaks that we wondered at the picnic life that had fallen to us. The village of Exeter was alive in all things social. The city families who had farms or country places near the village were so fond of them that they rarely closed them for more than two or three months, and these months were as likely to come in summer as in winter.

Our friends the Gordons made Homestead Farm their permanent residence, though they kept open house in town. Beyond the Gordons' was the modest home of an Irish baronet, Sir Thomas O'Hara. Sir Tom was a bachelor of sixty. He had run through two fortunes (as became an Irish baronet) in the racing field and at Homburg, and as a young man he had lived ten years at Limmer's tavern in London. When not in training to ride his own steeple-chasers, he was putting up his hands against any man in England who would face him for a few friendly rounds. He was not always victorious, either in the field, before the green cloth, or in the ring; but he was always a kind-hearted gentleman who would divide his last crown with friend or foe, and who could accept a beating with grace and unruffled spirit.

He could never ride below the welter weight, and after a few years he outgrew this weight and was forced to give up the least expensive of his diversions. The green cloth now received more of his attention, and, as a matter of course, of his money. Things went badly with him, and he began to see the end of his second fortune before he called a halt. Bad times in Ireland seriously reduced his rents, and he was forced to dispose of his salable estates. Then he came to this country in the hope of recouping himself, and to get away from the fast set that surrounded him.

"I can resist anything but temptation," this warm-hearted Irishman would say; and that was the keynote of his character.

Though Sir Tom was only sixty years old, he looked seventy. He was much broken in health by gout and the fast pace of his early manhood. But his spirit was untouched by misfortune, disease, or hardship. His courage was as good as when he served as a subaltern of the Guards in the trenches before Sebastopol, or presented his body as a mark for the sledge-hammer blows of Tom Sayers, just for diversion. His constitution must have been superb, for even in his decrepitude he was good to look upon: five feet ten, fine body, slightly given to rotundity, legs a little shrunken in the shanks, but giving unmistakable signs of what they had been ("not lost, but gone before," as he would say of them), hands and feet aristocratic in form and well cared for, and a fine head set on broad shoulders. His hair was thin, and he parted it with great exactness in the middle. His eyes were brown, large, and of exceeding softness. His nose was straight in spite of many a contusion, and his whole expression was that of a high-bred gentleman somewhat the worse for wear. Sir Tom was perfectly groomed when he came forth from his chamber, which was usually about ten in the morning.

Those of us who had access to his rooms often wondered how he ever got out of them looking so immaculate, for they were a perfectly impassable jungle to the stranger. Such a tangle of trunks, hand-bags, rug bundles, clothes, boots, pajamas, newspapers, scrap-books, B. & S. bottles, could hardly be found anywhere else in the world. He had a fondness for newspaper clippings, and had trunks of them, sorted into bundles or pasted in scrap-books. Old volumes of Bell's *Life* filled more than one trunk, and on one occasion when he and I were spending a long evening together, in celebration of his recent recovery from an attack of gout, and when he had done more than usual justice to the B. & S. bottles and less than usual justice to his gout, he showed me the record of a long-gone year in which this same Bell's *Life* called him the "first among the gentlemen riders in the United Kingdom," and proved this assertion by showing how he had won most of the great steeple-chases in England and Ireland, riding his own horses. This was the nearest approach to boasting that ever came to my knowledge in the years of our close friendship, and I would never have thought of it as such had I not seen that he regarded it as unwarrantable self-praise.

I have never known a more simple, kind-hearted, agreeable, and lovable gentleman than this broken-down sporting man and gambler. I loved him as a brother; and though he has passed out of my life, I still love the memory of his genial face, his courtesy, his unselfish friendship, more than words can express. A tender heart and a gentle spirit found strange housing in a body given over to reckless prodigality. The combination, tempered by time and exhaustion, showed nothing that was not lovable; and it is scant praise to say that Sir Thomas was much to me.

He was just as acceptable to Polly. No woman could fail to appreciate the homage which he never failed to show to the wife and mother. Many winter evenings at Four Oaks were made brighter by his presence, and we grew to expect him at least three nights each week. His plate was placed on our round table these nights, and he rarely failed to use it; and the B. & S. bottles were near at hand, and his favorite brand of cigars within easy reach.

"I light a 'baccy' by your permission, Mrs. Williams," and a courtly bow accompanied the words.

At 9.30 William came to bring Sir Tom home. The leave-taking was always formal with Polly, but with me it was, "Ta-ta, Williams—see you later," and our guest would hobble out on his poor crippled feet, waving his hand gallantly, with a voice as cheery as a boy's.

Another family whom I wish the reader to know well is the Kyrles. For more than twenty-five years we have known no joys or sorrows which they did not feel, and no interests that touched them have failed to leave a mark on us. We could not have been more intimate or better friends had the closest blood tie united us. The acquaintance of young married couples had grown into a friendship that was bearing its best fruit at a time when best fruit was most appreciated. We do not consider a pleasure more than half complete until we have told it to Will and Frances Kyrle, for their delight doubles our happiness.

They were among the earliest of my patients, and they are easily first among our friends. I have watched more than a half-dozen of their children from infancy to adult life, and this alone would be a strong bond; but in addition to this is the fact that the whole family, from father to youngest child, possess in a wonderful degree that subtle sense of true camaraderie which is as rare as it is charming.

The Kyrles lived in the city, but they were foot-free, and we could count on having them often. Four Oaks was to be, if we had our way, a country home for them almost as much as for us. Indeed, one of the rooms was called the Kyrles' room, and they came to it at will. Enough about our friends. We must go back to the farm interests, which are, indeed, the only excuse for this history.

CHAPTER XXXVII

The Headman's Job

Our life at Four Oaks began in earnest in January, 1897. Even during the winter months there was no lack of employment and interest for the Headman. I breakfasted at seven, and from that time until noon I was as busy as if I were working for $20 a month. The master's eye is worth more than his hand in a factory like mine. My men were, and are, an unusual lot,—intelligent, sober, and willing,— but they, like others, are apt to fall into routine ways, and thereby to miss points which an observing proprietor would not overlook.

The cows, for instance, were all fed the same ration. Fifteen pounds of mixed grains was none too much for the big Holstein milk-makers, who were yielding well and looking in perfect health; but the common cows were taking on too much flesh and falling off in milk. I at once changed the ration for these six cows by leaving out the corn entirely and substituting oat straw for alfalfa in the cut feed. The change brought good results in five of the cows; the other one did not pick up in her milk, and after a reasonable trial I sold her.

The herd was doing excellently for mid-winter,—the yield amounted to a daily average of 840 pounds throughout the month, and I was able to make good my contract with the middleman. I could see breakers ahead, however, and it behooved me to make ready for them. I decided to buy ten more thoroughbreds in new milk, if I could find them. I wrote to the people from whom I had purchased the first herd, and after a little delay secured nine cows in fresh milk and about four years old. This addition came in February, and kept my milk supply above the danger point. Since then I have bought no cows. Thirty-four of these thoroughbreds are still at Four Oaks—two of them have died, and three have been sold for not keeping up to the standard—and are doing grand service. Their numbers have been reënforced by twenty of their best daughters, so there are at this writing fifty-four milch cows and five yearling heifers in the herd. Most of the calves have been disposed of as soon as weaned. I have no room for more stock on my place, and it doesn't pay to keep them to sell as cows. Four Oaks is not a breeding farm, but a factory farm, and everything has to be subordinated to the factory idea.

My thoroughbred calves have brought me an average price of $12 each at four to six weeks, sold to dairymen, and I am satisfied to do business in that way. The nine milch cows which I bought to complete the herd cost, delivered at Four Oaks, $1012.

All the grain fed to cows, horses, and hogs, and a portion of that fed to chickens, is ground fine before feeding. The grinding is done in the granary by a mill with a capacity of forty bushels an hour. We make corn meal, corn and cob meal, and oatmeal enough for a week's supply in a few hours. All hay and straw is cut fine, before being fed, by a power cutter in the forage barn, and from thence is taken by teams in box racks to the feeding rooms, where it is wetted with hot water and mixed with the ground feed for the cows and horses, and steamed or cooked with the ground feed for the hogs and hens.

Alfalfa is the only hay used for the hens, and wonderfully good it is for them. Besides feed for the hogs, we have to provide ashes, salt, and charcoal for them. These three things are kept constantly before them in narrow troughs set so near the wall that they cannot get their feet into them.

We carefully save all wood ashes for the hogs and hens, and we burn our own charcoal in a pit in the wood lot. Five cords of sound wood make an abundant supply for a year. I think this side dish constantly before swine goes a long way toward keeping them healthy. Clean pens, well-balanced and well-cooked food, pure water, and this medicine can be counted on to keep a growing and fattening herd healthy during its nine months of life.

It is claimed that it is unnatural and artificial to confine these young things within such narrow limits, and so it is; but the whole scheme is unnatural, if you please. The pig is born to die, and to die quickly, for the profit and maintenance of man. What could be more unnatural? Would he be better reconciled to his fate after spending his nine months between field and sty? I wot not. The Chester White is an indolent fellow, and I suspect he loves his comfortable house, his cool stone porch, his back yard to dig in, his neighbors across the wire fence to gossip with, and his well-balanced, well-cooked food served under his own nose three times a day. At least he looks content in his piggery, and grows faster and puts on more flesh in his 250 days than does his neighbor of the field. If the hog's profitable life were twice or thrice as long, I would advocate a wider liberty for the early part of it; but as it doesn't pay to keep the animal after he is nine months old, the quickest way to bring him to perfection is the best. One cannot afford to graze animals of any kind when one is trying to do intensive farming. It is indirect, it is wasteful of space and energy, and it doesn't force the highest product. Grazing, as compared with soiling, may be economical of labor, but as I understand economics that is the one thing in which we do not wish to economize. The multiplication of well-paid and well-paying labor is a thing to be specially desired. If the soiling farm will keep two or three more men employed at good wages, and at the same time pay better interest than the grazing farm, it should be looked upon as much the better method. The question of furnishing landscape for hogs is one that borders too closely on the æsthetic or the sentimental to gain the approval of the factory-farm man. What is true of hogs is also true of cows. They are better off under the constant care of intelligent and interested human beings than when they follow the rippling brook or wind slowly o'er the lea at their own sweet pleasure.

The truth is, the rippling brook doesn't always furnish the best water, and the lea furnishes very imperfect forage during nine months of the year. A twenty-acre lot in good grass, in which to take the air, is all that a well-regulated herd of fifty cows needs. The clean, cool, calm stable is much to their liking, and the regular diet of a first-class cow-kitchen insures a uniform flow of milk.

What is true of hogs and cows is true also of hens. The common opinion that the farm-raised hen that has free range is healthier or happier than her sister in a well-ordered hennery is not based on facts. Freedom to forage for one's self and pick up a precarious living does not always mean health, happiness, or comfort. The strenuous life on the farm cannot compare in comfort with the quiet house and the freedom from anxiety of the well-tended hen. The vicissitudes of life are terrible for the uncooped chicken. The occupants of air, earth, and water lie in wait for it. It is fair game for the hawk and the owl; the fox, the weasel, the rat, the wood pussy, the cat, and the dog are its sworn enemies. The horse steps on it, the wheel crushes it; it falls into the cistern or the swill barrel; it is drenched by showers or stiffened by frosts, and, as the English say, it has a "rather indifferent time of it." If it survive the summer, and some chickens do, it will roost and shiver on the limb of an apple tree. Its nest will be accessible only to the mink and the rat; and, like Rachel, it will mourn for its children, which are not.

No, the well-yarded hen has by all odds the best of it. The wonder is that, with three-fourths of the poultry at large and making its own living, hens still furnish a product, in this country alone, $100,000,000 greater in value than the whole world's output of gold. Our annual production of eggs and poultry foots up to $280,000,000,—$4 apiece for every man, woman, and child,—and yet people say that hens do not pay!

Each flock of forty hens at Four Oaks has a house sixteen feet by twenty, and a run twenty feet by one hundred. I hear no complaints of close quarters or lack of freedom, but I do hear continually the song of contentment, and I see results daily that are more satisfactory than those of any oil well or mine in which I have ever been interested.

CHAPTER XXXVIII

Spring of '97

Sam began to make up his breeding pens in January. He selected 150 of his favorites, divided them into 10 flocks of 15, added a fine cockerel to each pen (we do not allow cocks or cockerels to run with the laying hens), and then began to set the incubator house in order.

He filled the first incubator on Saturday, January 30, and from that day until late in April he was able to start a fresh machine about every six days. Sam reports the total hatch for the year as 1917 chicks, out of which number he had, when he separated them in the early autumn, 678 pullets to put in the runs for laying hens, and 653 cockerels to go to the fattening pens. These figures show that Sam was a first-class chicken man.

We secured 300 tons of ice at the side of the lake for $98, having to pay a little more that year than the last, on account of the heavy fall of snow.

The wood-house was replenished, although there was still a good deal of last year's cut on hand. We did not fell any trees, for there was still a considerable quantity of dead wood on the ground which should be used first. I wanted to clear out much of the useless underbrush, but we had only time to make a beginning in this effort at forestry. We went over perhaps ten acres across the north line, removing briers and brush. Everything that looked like a possible future tree was left. Around oak and hickory stumps we found clumps of bushes springing from living roots. These we cut away, except one or possibly two of the most thrifty. We trimmed off the lower branches of those we saved, and left them to make such trees as they could. I have been amazed to see what a growth an oak-root sprout will make after its neighbors have been cut away. There are some hundreds of these trees in the forest at Four Oaks, from five to six inches in diameter, which did not measure more than one or two inches five years ago.

As the underbrush was cleared from the wood lot, I planned to set young trees to fill vacant spaces. The European larch was used in the first experiment. In the spring of 1897 I bought four thousand seedling larches for $80, planted them in nursery rows in the orchard, cultivated them for two years, and then transplanted them to the forest. The larch is hardy and grows rapidly; and as it is a valuable tree for many purposes, it is one of the best for forest planting. I have planted no others thus far at Four Oaks, as the four thousand from my little nursery seem to fill all unoccupied spaces.

Fresh mulching was piled near all the young fruit trees, to be applied as soon as the frost was out of the ground. Several hundreds of loads of manure were hauled to the fields, to be spread as soon as the snow disappeared. I always return manure to the land as soon as it can be done conveniently. The manure from the hen-house was saved this year to use on the alfalfa fields, to see how well it would take the place of commercial fertilizer. I may as well give the result of the experiment now.

It was mixed with sand and applied at the rate of eight hundred pounds an acre for the spring dressing over a portion of the alfalfa, against four hundred pounds an acre of the fertilizer 3:8:8. After two years I was convinced that, when used alone, it is not of more than half the value of the fertilizer.

My present practice is to use five hundred pounds of hen manure and two hundred pounds of fertilizer on each acre for the spring dressing, and two hundred pounds an acre of the fertilizer alone after each cutting except the last. We have ten or twelve tons of hen manure each year, and it is nearly all used on the alfalfa or the timothy as spring dressing. It costs nothing, and it takes off a considerable sum from the fertilizer account. I am not at all sure that the scientists would approve this method of using it; I can only give my experience, and say that it brings me satisfactory crops.

There was much snow in January and February, and in March much rain. When the spring opened, therefore, the ground was full of water. This was fortunate, for April and May were unusually dry months,—only 1.16 inches of water.

The dry April brought the ploughs out early; but before we put our hands to the plough we should make a note of what the first quarter of 1897 brought into our strong box.

Sold:

Butter	$842.00
Eggs	401.00
Cow	35.00
Two sows	19.00
Total	$1297.00

Fifteen of the young sows farrowed in March, and the other 9 in April, as also did 18 old ones. The young sows gave us 147 pigs, and the old ones 161, so that the spring opened with an addition to our stock of 300 head of young swine.

Between March 1 and May 10 were born 25 calves, which were all sold before July 1. The population of our factory farm was increasing so rapidly that it became necessary to have more help. We already had eight men and three women, besides the help in the big house. One would think that eight men could do the work on a farm of 320 acres, and so they can, most of the time; but in seed-time and harvest they are not sufficient at Four Oaks. We could not work the teams.

Up to March, 1897, Sam had full charge of the chickens, and also looked after the hogs, with the help of Anderson. Judson and French had their hands full in the cow stables, and Lars was more than busy with the carriage horses and the driving. Thompson was working foreman, and his son Zeb and Johnson looked after the farm horses during the winter and did the general work. From that time on Sam gave his entire time to the chickens, Anderson his entire time to the hogs, and Johnson began gardening in real earnest. This left only Thompson and Zeb for general farm work.

Again I advertised for two farm hands. I selected two of the most promising applicants and brought them out to the farm. Thompson discharged one of them at the end of the first day for persistently jerking his team, and the other discharged himself at the week's end, to continue his tramp. Once more I resorted to the city papers. This time I was more fortunate, for I found a young Swede, square-built and blond-headed, who said he had worked on his father's farm in the old country, and had left it because it was too small for the five boys. Otto was slow of speech and of motion, but he said he could work, and I hired him. The other man whom I sent to the farm at the same time proved of no use whatever. He stayed four days, and was dismissed for innocuous desuetude. Still another man whom I tried did well for five weeks, and then broke out in a most profound spree, from which he could not be weaned. He ended up by an assault on Otto in the stable yard. The Swede was taken by surprise, and was handsomely bowled over by the first onslaught of his half-drunk, half-crazed antagonist. As soon, however, as his slow mind took in the fact that he was being pounded, he gathered his forces, and, with a grunt for a war-cry, rolled his enemy under him, sat upon his stomach, and, flat-handed, slapped his face until he shouted for aid. The man left the farm at once, and I commended the Swede for having used the flat of his hand.

In spite of bad luck with the new men we were able to plough and seed 144 acres by May 10. Lots Nos. 8, 12, 13, and 14 were planted to corn, and No. 15 sowed to oats, and the 10 acres on the home lot were divided between sweet fodder corn, potatoes, and cabbage. The abundant water in the soil gave the crops a fair start, and June proved an excellent growing month, a rainfall of nearly four inches putting them beyond danger from the short water supply of July and August. Indeed, had it not been for the generosity of June we should have been in a bad way, for the next three months gave a scant four inches of rain.

The oats made a good growth, though the straw was rather short, and the corn did very well indeed,—due largely to thorough cultivation. Twelve acres of oats were cut for forage, and the rest yielded 33 bushels to the acre,—a little over 1300 bushels.

The alfalfa and timothy made a good start. From the former we cut, late in June, 2¼ tons to the acre, and from the timothy, in July, 2½ tons,—50 tons of timothy and 45 of alfalfa. Each of these fields received the usual top-dressing after the crop was cut; but the timothy did not respond,—the late season was too dry. We cut two more crops from the alfalfa field, which together made a yield of a little more than 2 tons. The alfalfa in that dry summer gave me 95 tons of good hay, proving its superiority as a dry-weather crop.

Johnson started the one-and-one-half-acre vegetable and fruit garden in April, and devoted much of his time to it. His primitive hotbeds gradually emptied themselves into the garden, and we now began to taste the fruit of our own soil, much to the pleasure of the whole colony. It is surprising what a real gardener can do with a garden of this size. By feeding soil and plants liberally, he is able to keep the ground producing successive crops of vegetables, from the day the frost leaves it in the spring until it again takes possession in the fall, without doing any wrong to the land. Indeed, our garden grows better and more prolific each year in spite of the immense crops that are taken from it. This can be done only by a person who knows his business, and Johnson is such a person. He gave much of his time to this practical patch, but he also worked with Polly among the shrubs on the lawn, and in her sunken flower garden, which is the pride of her life. We shall hear more about this flower garden later on.

The accounts for the second quarter of the year show these items on the income side:—

Butter. $1052.00
Eggs. 379.00
Twenty-five calves 275.00
Total $1706.00

CHAPTER XXXIX

The Young Orchard

One of the most enjoyable occupations of a farmer's life is the care of young trees. Until your experience in this work is of a personal and proprietary nature, you will not realize the pleasure it can afford. The intimate study of plant life, especially if that plant life is yours, is a never failing source of pleasurable speculation, and a thing upon which to hang dreams. You grow to know each tree, not only by its shape and its habit of growth, but also by peculiarities that belong to it as an individual. The erect, sturdy bearing of one bespeaks a frank, bold nature, which makes it willing to accept its surroundings and make the most of them; while the crooked, dwarfish nature of another requires the utmost care of the husbandman to keep it within the bounds of good behavior. And yet we often find that the slow-growing, ill-conditioned young tree, if properly cared for, will bring forth the finest fruit at maturity.

To study the character and to watch the development of young trees is a pleasing and useful occupation for the man who thinks of them as living things with an inheritance that cannot be ignored. That seeds in all appearance exactly alike should send forth shoots so unlike, is a wonder of Nature; and that young shoots in the same soil and with the same care should show such dissimilarity in development, is a riddle whose answer is to be found only in the binding laws of heredity. That a tiny bud inserted under the bark of a well-grown tree can change a sour root to a sweet bough, ought to make one careful of the buds which one grafts on the living trunk of one's tree of life. The young orchard can teach many lessons to him who is willing to be taught; in the hands of him who is not, the schoolmaster has a very sorry time of it, no matter how he sets his lessons.

The side pockets of my jacket are usually weighted down with pruning-shears, a sharp knife, and a handled copper wire,—always, indeed, in June, when I walk in my orchard. June is the month of all months for the prudent orchardist to go thus armed, for the apple-tree borer is abroad in the land. When the quick eye of the master sees a little pile of sawdust at the base of a tree, he knows that it is time for him to sit right down by that tree and kill its enemy. The sharp knife enlarges the hole, which is the trail of the serpent, and the sharp-pointed, flexible wire follows the route until it has reached and trans-fixed the borer.

This is the only way. It is the nature of the borer to maim or kill the tree; it is for the interest of the owner that the tree should live. The conflict is irrepressible, and the weakest must go to the wall. The borer evil can be reduced to a minimum by keeping the young trees banked three or four inches high with firm dirt or ashes; but borers must be followed with the wire, once they enter the bark.

The sharp knife and the pruning-shears have other uses in the June orchard. Limbs and sprouts will come in irregular and improper places, and they should be nipped out early and thus save labor and mutilation later on. Sprouts that start from the eyes on the trunk can be removed by a downward stroke of the gloved hand. All inter-secting or crossing boughs are removed by knife or scissors, and branches which are too luxuriant in growth are cut or pinched back. Careful guidance of the tree in June will avoid the necessity of severe correction later on.

A man ought to plant an orchard, if for no other reason, that he may have the pleasure of caring for it, and for the companionship of the trees. This was the second year of growth for my orchard, and I was gratified by the evidences of thrift and vigor. Fine, spreading heads adorned the tops of the stubs of trees that had received such (apparently) cruel treatment eighteen months before. The growth of these two seasons convinced me that the four-year-old root and the three-year-old stem, if properly managed, have greater possibilities of rapid development than roots or stems of more tender age. I think I made no mistake in planting three-year-old trees.

As I worked in my orchard I could not help looking forward to the time when the trees would return a hundred-fold for the care bestowed upon them. They would begin to bring returns, in a small way, from the fourth year, and after that the returns would increase rapidly. It is safe to predict that from the tenth to the fortieth year a well-managed orchard will give an average yearly income of $100 an acre above all expenses, including interest on the original cost. A fifty-acre orchard of well-selected apple trees, near a first-class market and in intelligent hands, means a net income of $5000, taking one year with another, for thirty or forty years. What kind of investment will pay better? What sort of business will give larger returns in health and pleasure?

I do not mean to convey the idea that forty years is the life of an orchard; hundreds of years would be more correct. As trees die from accident or decrepitude, others should take their places. Thus the lease of life becomes perpetual in hands that are willing to keep adding to the soil more than the trees and the fruit take from it. Comparatively few owners of orchards do this, and those who belong to the majority will find fault with my figures; but the thinking few, who do not expect to enjoy the fat of the land without making a reasonable return, will say that I am too conservative,—that a well-placed, well-cared-for, well-selected, and well-marketed orchard will do much better than my prophecy. Nature is a good husbandman so far as she goes, but her scheme contemplates only the perpetuation of the tree, by seeds or by other means. Nature's plan is to give to each specimen a nutritive ration. Anything beyond this is thrown away on the individual, and had better be used for the multiplying of specimens. When man comes to ask something more than germinating seeds from a plant, he must remove it from the crowded clump, give it more light and air, *and feed it for product*. In other words, he must give it more nitrogen, phosphoric acid, and potash than it can use for simple growth and maintenance, and thus make it burst forth into flower-or fruit-product. Nature produces the apple tree, but man must cultivate it and feed it if he would be fed and comforted by it. People who neglect their orchards can get neither pleasure nor profit from them, and such persons are not competent to sit in judgment upon the value of an apple tree. Only those who love, nourish, and profit by their orchards may come into the apple court and speak with authority.

CHAPTER XL

The Timothy Harvest

On Friday, the 25th, the children came home from their schools, and with them came Jim Jarvis to spend the summer holidays. Our invitation to Jarvis had been unanimous when he bade us good-by in the winter. Jack was his chum, Polly had adopted him, I took to him from the first, and Jane, in her shy way, admired him greatly. The boys took to farm life like ducks to water. They were hot for any kind of work, and hot, too, from all kinds. I could not offer anything congenial until the timothy harvest in July. When this was on, they were happy and useful at the same time,—a rare combination for boys.

The timothy harvest is attractive to all, and it would be hard to find a form of labor which contributes more to the æsthetic sense than does the gathering of this fragrant grass. At four o'clock on a fine morning, with the barometer "set fair," Thompson started the mower, and kept it humming until 6.30, when Zeb, with a fresh team, relieved him. Zeb tried to cut a little faster than his father, but he was allowed no more time. Promptly at nine he was called in, and there was to be no more cutting that day. At eleven o'clock the tedder was started, and in two hours the cut grass had been turned. At three o'clock the rake gathered it into windrows, from which it was rolled and piled into heaps, or cocks, of six hundred or eight hundred pounds each. The cutting of the morning was in safe bunches before the dew fell, there to go through the process of sweating until ten o'clock the next day. It was then opened and fluffed out for four hours, after which all hands and all teams turned to and hauled it into the forage barn.

The grass that was cut one morning was safely housed as hay by the second night, if the weather was favorable; if not, it took little harm in the haycocks, even from foul weather. It is the sun-bleach that takes the life out of hay.

This year we had no trouble in getting fifty tons of as fine timothy hay as horses could wish to eat or man could wish to see. We began to cut on Tuesday, the 6th of July, and by Saturday evening the twenty-acre crop was under cover. The boys blistered their hands with the fork handles, and their faces, necks, and arms with the sun's rays, and claimed to like the work and the blisters. Indeed, tossing clean, fragrant hay is work fit for a prince; and a man never looks to better advantage or more picturesque than when, redolent with its perfume, he slings a jug over the crook in his elbow and listens to the gurgle of the home-made ginger ale as it changes from jug to throat. There may be joys in other drinks, but for solid comfort and refreshment give me a July hay-field at 3 P.M., a jug of water at forty-eight degrees, with just the amount of molasses, vinegar, and ginger that is Polly's secret, and I will give cards and spades to the broadest goblet of bubbles that was ever poured, and beat it to a standstill. Add to this a blond head under a broad hat, a thin white gown, such as grasshoppers love, and you can see why the emptying of the jug was a satisfying function in our field; for Jane was the one who presided at these afternoon teas. Often Jane was not alone; Florence or Jessie, or both, or others, made hay while the sun shone in those July days, and many a load went to the barn capped with white and laughter. The young people decided that a hay farm would be ideal—no end better than a factory farm—and advised me to put all the land into timothy and clover. I was not too old to see the beauties of haying-time, with such voluntary labor; but I was too old and too much interested with my experiment to be cajoled by a lot of youngsters. I promised them a week of haying in each fifty-two, but that was all the concession I would make. Laura said:—

"We are commanded to make hay while the sun shines; and the sun always shines at Four Oaks, for me."

It was pretty of her to say that; but what else would one expect from Laura?

The twelve acres from which the fodder oats had been cut were ploughed and fitted for sugar beets and turnips. I was not at all certain that the beets would do anything if sown so late, but I was going to try. Of the turnips I could feel more certain, for doth not the poet say:—

> "The 25th day of July,
> Sow your turnips, wet or dry"?

As the 25th fell on Sunday, I tried to placate the agricultural poet by sowing half on the 24th and the other half on the 26th, but it was no use. Whether the turnip god was offended by the fractured rule and refused his blessing, or whether the dry August and September prevented full returns, is more than I can say. Certain it is that I had but a half crop of turnips and a beggarly batch of beets to comfort me and the hogs.

Some little consolation, however, was found in Polly's joy over a small crop of currants which her yearling bushes produced. I also heard rumors of a few cherries which turned their red cheeks to the sun for one happy day, and then disappeared. Cock Robin's breast was red the next morning, and on this circumstantial evidence Polly accused him. He pleaded "not guilty," and strutted on the lawn with his thumbs in the armholes of his waistcoat and his suspected breast as much in evidence as a pouter pigeon's. A jury, mostly of blackbirds, found the charge "not proven," and the case was dismissed. I was convinced by the result of this trial that the only safe way would be to provide enough cherries for the birds and for the people too, and ordered fifty more trees for fall planting. I found by experience, that if one would have bird neighbors (and who would not?), he must provide liberally for their wants and also for their luxuries. I have stolen a march as to the cherries by planting scores of mulberry trees, both native and Russian. Birds love mulberries even better than they do cherries, and we now eat our pies in peace. To make amends for this ruse, I have established a number of drinking fountains and free baths; all of which have helped to make us friends.

In August I sold, near the top of a low market, 156 young hogs. At $4.50 per hundred, the bunch netted me $1807. They did not weigh quite as much as those sold the previous autumn, and I found two ways of accounting for this. The first and most probable was that fall pigs do not grow so fast as those farrowed in the spring. This is sufficient to account for the fact that the herd average was twenty pounds lighter than that of its predecessor. I could not, however, get over the notion that Anderson's nervousness had in some way taken possession of the swine (we have Holy Writ for a similar case), and that they were wasted in growth by his spirit of unrest. He was uniformly kind to them and faithful with their food, but there was lacking that sense of cordial sympathy which should exist between hog and man if both would appear at their best. Even when Anderson came to their pens reeking with the rich savor of the food they loved, their ears would prick up (as much as a Chester White's ears can), and with a "woof!" they would shoot out the door, only to return in a moment with the greatest confidence. I never heard that "woof" and saw the stampede without looking around for the "steep place" and the "sea," feeling sure that the incident lacked only these accessories to make it a catastrophe.

Anderson was good and faithful, and he would work his arms and legs off for the pigs; but the spirit of unrest entered every herd which he kept, though neither he nor I saw it clearly enough to go and "tell it in the city." With other swineherds my hogs averaged from fifteen to eighteen pounds better than with faithful Anderson, and I am, therefore, competent to speak of the gross weight of the spirit of contentment.

CHAPTER XLI

Strike at Gordon's Mine

Frank Gordon owned a coal mine about six miles west of the village of Exeter, and four miles from Four Oaks. A village called Gordonville had sprung up at the mouth of the mine. It was the home of the three hundred miners and their families,—mostly Huns, but with a sprinkling of Cornishmen.

The houses were built by the owner of the mine, and were leased to the miners at a small yearly rental. They were modest in structure, but they could be made inviting and neat if the occupants were thrifty. No one was allowed to sell liquor on the property owned by the Gordons, but outside of this limit was a fringe of low saloons which did a thriving business off the improvident miners.

There had never been a strike at Gordonville, and such a thing seemed improbable, for Gordon was a kind master, who paid his men promptly and looked after their interests more than is usual for a capitalist.

It was, therefore, a distinct surprise when the foreman of the mine telephoned to Gordon one July morning that the men had struck work. Gordon did not understand the reason of it, but he expressed himself as being heartily glad, for financial reasons, that the men had gone out. He had more than enough coal on the surface and in cars to supply the demand for the next three months, and it would be money in his pocket to dispose of his coal without having to pay for the labor of replacing it.

During the day the reason for the strike was announced. From the establishment of the mine it had been the custom for the miners to have their tools sharpened at a shop built and run by the property. This was done for the accommodation of the men, and the charge for keeping the tools sharp was ten cents a week for each man, or $5 a year. For twenty years no fault had been found with the arrangement; it had been looked upon as satisfactory, especially by the men. A walking delegate, mousing around the mine, and finding no other cause for complaint, had lighted upon this practice, and he told the men it was a shame that they should have to pay ten cents a week out of their hard-earned wages for keeping their tools sharp. He said that it was the business of the property to keep the tools sharp, and that the men should not be called upon to pay for that service; that they ought, in justice to themselves and for the dignity of associated labor, to demand that this onerous tax be removed; and, to insure its removal, he declared a strike on. This was the reason, and the only reason, for the strike at Gordon's mine. Three hundred men quit work, and three hundred families suffered, many of them for the necessities of life, simply because a loud-mouthed delegate assured them that they were being imposed upon.

Things went on quietly at the mine. There was no riot, no disturbance. Gordon did not go over, but simply telephoned to the superintendent to close the shaft houses, shut down the engines, put out the fires, and let things rest, at the same time saying that he would hold the superintendent and the bosses responsible for the safety of the plant.

The men were disappointed, as the days went by, that the owner made no effort to induce them to resume work. They had believed that he would at once accede to their demand, and that they would go back to work with the tax removed. This, however, was not his plan. Weeks passed and the men became restless. They frequented the saloons more generally, spent their remaining money for liquor, and went into debt as much as they were permitted for more liquor. They became noisy and quarrelsome. The few men who were opposed to the strike could make no headway against public opinion. These men held aloof from the saloons, husbanded their money, and confined themselves as much as possible to their own houses.

Things had gone on in this way for six weeks. The men grew more and more restless and more dissipated. Again the walking delegate came to encourage them to hold out. Mounted on an empty coal car, he made an inflammatory speech to the men, advising them not only to hold out against the owner, but also to prevent the employment of any other help. If this should not prove sufficient, he advised them to wreck the mining property and to fire the mine,—to bring the owner to terms.

Jack and Jarvis went for a long walk one day, and their route took them near Gordonville. Seeing the men collected in such numbers around a coal car, they approached, and heard the last half of this inflammatory speech. As the walking delegate finished, Jack jumped up on the car, and said:—

"McGinnis has had his say; now, men, let me have mine. There are always two sides to a question. You have heard one, let me give you the other. I am a delegate, self-appointed, from the amalgamated Order of Thinkers, and I want you to listen to our view of this strike,—and of all strikes. I want you also to think a little as well as to listen.

"You have been led into this position by a man whose sole business is to foment discords between working-men and their employers. The moment these discords cease, that moment this man loses his job and must work or starve like the rest of you. He is, therefore, an interested party, and he is more than likely to be biassed by what seems to be his interest. He has made no argument; he has simply asserted things which are not true, and played upon your sympathies, emotions, and passions, by the use of the stale war-cries—'oppression,' 'down-trodden working-man,' 'bloated bondholders,' and, most foolish of all, 'the conflict between Capital and Labor.' You have not thought this matter out for yourselves at all. That is why I ask you to join hands for a little while with the Order of Thinkers and see if there is not some good way out of this dilemma. McGinnis said that the Company has no right to charge you for keeping your tools sharp. In one sense this is true. You have a perfect right to work with dull tools, if you wish to; you have the right to sharpen your own tools; and you also have the right to hire any one else to do it for you. You work 'by the ton,' you own your pickaxes and shovels from handle to blade, and you have the right to do with them as you please.

"There are three hundred of you who use tools; you each pay ten cents a week to the Company for keeping them sharp,—that is, in round numbers, $1500 a year. There are two smiths at work at $50 a month (that is $1200), and a helper at $25 a month ($300 more), making just $1500 paid by the Company in wages. If you will think this matter out, you will see that there is a dead loss to the Company of the coal used, the wear and tear of the instruments, and the interest, taxes, insurance, and degeneration of the plant. Is the Company under obligation to lose this money for you? Not at all! The Company does this as an accommodation and a gratuity to you, but not as a duty. Just as much coal would be taken from the Gordon mine if your tools were never sharpened, only it would require more men, and you would earn less money apiece. You could not get this sharpening done at private shops so cheaply, and you cannot do it yourselves. You have no more right to ask the Company to do this work for nothing than you have to ask it to buy your tools for you. It would be just as sensible for you to strike because the Company did not send each of you ten cents' worth of ice-cream every Sunday morning, as it is for you to go out on this matter of sharpening tools.

"But, suppose the Company were in duty bound to do this thing for you, and suppose it should refuse; would that be a good reason for quitting work? Not by any means! You are earning an average of $2 a day,—nearly $16,000 a month. You've 'been out' six weeks. If you gain your point, it will take you fifteen years to make up what you've already lost. If you have the sense which God gives geese, you will see that you can't afford this sort of thing.

"But the end is not yet. You are likely to stay out six weeks longer, and each six weeks adds another fifteen years to your struggle to catch up with your losses. Is this a load which thinking people would impose upon themselves? Not much! You will lose your battle, for your strike is badly timed. It seems to be the fate of strikes to be badly timed; they usually occur when, on account of hard times or over-supply, the employers would rather stop paying wages than not. That's the case now. Four months of coal is in yards or on cars, and it's an absolute benefit to the Company to turn seventy or eighty thousand dollars of dead product into live money. Don't deceive yourselves with the hope that you are distressing the owner by your foolish strike; you are putting money into his pockets while your families suffer for food. There is no great principle at stake to make your conduct seem noble and to call forth sympathy for your suffering,—only foolishness and the blind following of a demagogue whose living depends upon your folly.

"McGinnis talked to you about the conflict between capital and labor. That is all rot. There is not and there cannot be such a conflict.

Labor makes capital, and without capital there would be no object in labor. They are mutually dependent upon each other, and there can be no quarrel between them, for neither could exist after the death of the other. The capitalist is only a laborer who has saved a part of his wages, —either in his generation or in some preceding one. Any man with a sound mind and a sound body can become a capitalist. When the laborer has saved one dollar he is a capitalist,—he has money to lend at interest or to invest in something that will bring a return. The second dollar is easier saved than the first, and every dollar saved is earning something on its own account. All persons who have money to invest or to lend are capitalists. Of course, some are great and some are small, but all are independent, for they have more than they need for immediate personal use.

"I am going to tell you how you may all become capitalists; but first I want to point out your real enemies. The employer is not your enemy, capital is not your enemy, but the saloonkeeper is,—and the most deadly enemy you can possibly have. In that fringe of shanties over yonder live the powers that keep you down; there are the foes that degrade you and your families, forcing you to live little better than wild beasts. Your food is poor, your clothing is in rags, your children are without shoes, your homes are desolate, there are no schools and no social life. Year follows year in dreary monotone, and you finally die, and your neighbors thrust you underground and have an end of you. Misery and wretchedness fill the measure of your days, and you are forgotten.

"This dull, brutish condition is self-imposed, and to what end? That some dozen harpies may fatten on your flesh; that your labor may give them leisure; that your suffering may give them pleasure; that your sweat may cool their brows, and your money fill their tills!

"What do you get in return? Whiskey, to poison your bodies and pervert your minds; whiskey, to make you fierce beasts or dull brutes; whiskey, to make your eyes red and your hands unsteady; whiskey, to make your homes sties and yourselves fit occupants for them; whiskey, to make you beat your wives and children; whiskey, to cast you into the gutter, the most loathsome animal in all the world. This is cheap whiskey, but it costs you dear. All that makes life worth living, all that raises man above the brute, and all the hope of a future life, are freely given for this poor whiskey. The man who sells it to you robs you of your money and also of your manhood. You pay him ten times (often twenty times) as much as it cost him, and yet he poses as your friend.

"I'm not going to say anything against beer, for I don't think good beer is very likely to hurt a man. I will say this, however,—you pay more than twice what it is worth. This is the point I would make: beer is a food of some value, and it should be put on a food basis in price. It isn't more than half as valuable as milk, and it shouldn't cost more than half as much. You can have good beer at three or four cents a quart, if you will let whiskey alone.

"I promised to tell you how to become capitalists, each and every one of you, and I'll keep my word if you'll listen to me a little longer."

While Jack had been speaking, some of the men had shown considerable interest and had gradually crowded their way nearer to the boy. Thirty or forty Cornishmen and perhaps as many others of the better sort were close to the car, and seemed anxious to hear what he had to say. Back of these, however, were the large majority of the miners and the hangers-on at the saloons, who did not wish to hear, and did not mean that others should hear, what the boy had to say. Led by McGinnis and the saloon-keepers, they had kept up such a row that it had been impossible for any one, except those quite near the car, to hear at all. Now they determined to stop the talk and to bounce the boy. They made a vigorous rush for the car with shouts and uplifted hands.

A gigantic Cornishman mounted the car, and said, in a voice that could easily be heard above the shouting of the crowd:—

"Wait—wait a bit, men! The lad is a brave one, and ye maun own to that! There be small 'urt in words, and mebbe 'e 'ave tole a bit truth. Me and me mates 'ere are minded to give un a chance. If ye men don't want to 'ear 'im, you don't 'ave to stay; but don't 'e dare touchen with a finger, or, by God! Tom Carkeek will kick the stuffin' out en 'e!"

This was enough to prevent any overt act, for Tom Carkeek was the champion wrestler in all that county; he was fiercer than fire when roused, and he would be backed by every Cornishman on the job.

Jack went on with his talk. "The 'Order of Thinkers' claim that you men and all of your class spend one-third of your entire wages for whiskey and beer. There are exceptions, but the figures will hold good. I am going to call the amount of your wages spent in this way, one-fourth. The yearly pay-roll of this mine is, in round numbers, $200,000. Fifty thousand of this goes into the hands of those harpies, who grow rich as you grow poor. You are surprised at these figures, and yet they are too small. I counted the saloons over there, and I find there are eleven of them. Divide $50,000 into eleven parts, and you would give each saloon less than $5000 a year as a gross business. Not one of those places can run on the legitimate percentage of a business which does not amount to more than that. Do you suppose these men are here from charitable motives or for their health? Not at all. They are here to make money, and they do it. Five or six hundred dollars is all they pay for the vile stuff for which they charge you $5000. They rob you of manhood and money alike.

"Now, what would be the result if you struck on these robbers? I will tell you. In the first place, you would save $50,000 each year, and you would be better men in every way for so doing. You would earn more money, and your children would wear shoes and go to school. That would be much, and well worth while; but that is not the best of it. I will make a proposition to you, and I will promise that it shall be carried out on my side exactly as I state it.

"This is a noble property. In ten years it has paid its owner $500,000,—$50,000 a year. It is sure to go on in this way under good management. I offer, in the name of the owner, to bond this property to you for $300,000 for five years at six per cent. Of course this is an unusual opportunity. The owner has grown rich out of it, and he is now willing to retire and give others a chance. His offer to you is to sell the mine for half its value, and, at the same time, to give you five years in which to pay for it. I will add something to this proposition, for I feel certain that he will agree to it. It is this: Mr. Gordon will build and equip a small brewery on this property, in which good, wholesome beer can be made for you at one cent a glass. You are to pay for the brewery in the same way that you pay for the other property; it will cost $25,000. This will make $325,000 which you are to pay during the next five years. How? Let me tell you.

"The property will give you a net income of $40,000 or $50,000, and you will save $50,000 more when you give up whiskey and get your beer for less than one-fourth of what it now costs you. The general store at which you have always traded will be run in your interests, and all that you buy will be cheaper. The market will be a cooperative one, which will furnish you meat, fattened on your own land, at the lowest price. Your fruit and vegetables will come from these broad acres, which will be yours and will cost you but little. You will earn more money because you will be sober and industrious, and your money will purchase more because you will deal without a middleman. You will be better clothed, better fed, and better men. Your wives will take new interest in life, and there will be carpets on your floors, curtains at your windows, vegetables behind your cottages, and flowers in front of them.

"All these things you will have with the money you are now earning, and at the same time you will be changing from the laborer to the capitalist. The mine gives you a profit of $40,000, and you save one-fourth of your wages, which makes $50,000 more,—$90,000 in all. What are you to do with this? Less than $20,000 will cover the interest. You will have $70,000 to pay on the principal. This will reduce the interest for the next year more than $3000. Each year you can do as well, and by the time the five years have passed you will own the mine, the land, the brewery, the store, the market, and this blessed blacksmith shop about which you have had so much fuss, and also a bank with a paid-up capital of $50,000. You are capitalists, every one of you, at the end of five years, if you wish to be, and if you are willing to give up the single item,—whiskey.

"Do you like the plan? Do you like the prospect? Turn it over and see what objections you can find. If you are willing to go into it, come over to Four Oaks some day and we will go more into details. McGinnis gave you one side of the picture: I have given you the other. You are at liberty to follow whichever you please."

Jack and Jarvis jumped off the car and struck out for home. Carkeek and his Cornishmen followed the lads until they were well clear of the village, to protect them, and then Carkeek said:—"Me and the others like for to hear 'e talk, mister, and we like for to 'ear 'e talk more."

"All right, Goliath," said Jack. "Come over any time and we'll make plans."

CHAPTER XLII

The Riot

Two days later the boys, returning from the city, were met by Jane and Jessie in the big carriage to be driven home. Halfway to Four Oaks the carriage suddenly halted, and a confused murmur of angry voices gave warning of trouble. Jack opened the door and stood upon the step.

"Fifteen or twenty drunken miners block the way,—they are holding the horses," said he.

"Let me out; I'll soon clear the road," said Jarvis, trying to force his way past Jack.

"Sit still, Hercules; I am slower to wrath than you are. Let me talk to them," and Jack took three or four steps forward, followed closely by Jarvis.

"Well, men, what do you want? There is no good in stopping a carriage on the highroad."

"We want work and money and bread," said a great bearded Hun who was nearest to Jack.

"This is no way to get either. We have no work to offer, there is no bread in the carriage, and not much money. You are dead wrong in this business, and you are likely to get into trouble. I can make some allowance when I remember the bad whiskey that is in you, but you must get out of our way; the road is public and we have the right to use it."

"Not until you have paid toll," said the Hun.

"That's the rooster who said we drank whiskey and didn't work. He's the fellow who would rob a poor man of his liberty," came a voice in the crowd.

"Knock his block off!"

"Break his back!"

"Let me at him," and a score of other friendly offers came from the drunken crowd.

Jack stood steadily looking at the ruffians, his blue eyes growing black with excitement and his hands clenched tightly in the pockets of his reefer.

"Slowly, men, slowly," said he. "If you want me, you may have me. There are ladies in the carriage; let them go on; I'll stay with you as long as you like. You are brave men, and you have no quarrel with ladies."

"Ladies, eh!" said the Hun, "ladies! I never saw anything but *women*. Let's have a look at them, boys."

This speech was drunkenly approved, and the men pressed forward. Jack stood firm, his face was white, but his eyes flamed.

"Stand off! There are good men who will die for those ladies, and it will go hard but bad men shall die first."

The Hun disregarded the warning.

"I'll have a look into—"

"Hell!" said the slow-of-wrath Jack, and his fist went straight from the shoulder and smote the Hun on the point of the jaw. It was a terrible blow, dealt with all the force of a trained athlete, and inspired by every impulse which a man holds dear; and the half-drunken brute fell like a stricken ox. Catching the club from the falling man, Jack made a sudden lunge forward at the face of the nearest foe.

"Now, Jim!" he shouted, as the full fever of battle seized him. His forward lunge had placed another miner *hors de combat*, and Jarvis sprang forward and secured the wounded man's bludgeon.

"Back to back, Jack, and mind your guard!"

The odds were eighteen to two against the young men, but they did not heed them. Back to back they stood, and the heavy clubs were like feathers in their strong hands. Their skill at "single stick" was of immense advantage, for it built a wall of defence around them. The crazy-drunk miners rushed upon them with the fierceness of wild beasts; they crowded in so close as to interfere with their own freedom of movement; they sought to overpower the two men by weight of numbers and by showers of blows. Jack and Jim were kept busy guarding their own heads, and it was only occasionally that they could give an aggressive blow. When these opportunities came, they were accepted with fierce delight, and a miner fell with a broken head at every blow. Two fell in front of Jack and three went down under Jarvis's club. The battle had now lasted several minutes, and the strain on the young men was telling on their wind; they struck as hard and parried as well as at first, but they were breathing rapidly. The young men cheered each other with joyous words; they felt no need of aid.

"Beats football hollow!" panted Jarvis.

"Go in, old man! you're a dandy full-back!" came between strokes from Jack.

Let us leave the boys for a minute and see what the girls are doing. When Jarvis got out of the carriage, he said:—

"Lars, if there is trouble here, you drive on as soon as you can get your horses clear. Never mind us; we'll walk home. Get the ladies to Four Oaks as soon as possible."

When the battle began, the miners left the horses to attack the men. This gave a clear road, and Lars was ready to drive on, but the girls were not in the carriage. They had sprung out in the excitement of the first sound of blows; and now stood watching with glowing eyes and white faces the prowess of their champions. For minutes they watched the conflict with fear and pride combined. When seven or eight minutes had passed and the champions had not slain all their enemies, some degree of terror arose in the minds of the young ladies,—terror lest their knights be overpowered by numbers or become exhausted by slaying,—and they looked about for aid. Lars, remembering what Jarvis had said, urged the ladies to get into the carriage and be driven out of danger. They repelled his advice with scorn. Jane said:—"I won't stir a step until the men can go with us!"

Jessie said never a word, but she darted forward toward the fighting men, stooped, picked up a fallen club, and was back in an instant. Mounting quickly to the box, she said:—"I can hold the horses. Don't you think you can help the men, Lars?"

"I'd like to try, miss," and the coachman's coat was off in a trice and the club in his hand. He was none too soon!

Jane, who had mounted the box with Jessie, cried, "Look out, Jack!" just as a heavy stone crashed against the back of his head. Some brute in the crowd had sent it with all his force. The stone broke through the Derby hat and opened a wide gash in Jack's scalp, and sent him to the ground with a thousand stars glittering before his eyes. Jane gave a sob and covered her eyes. Jessie swayed as though she would fall, but she never took her eyes from the fallen man; her lips moved, but she said nothing; and her face was ghastly white. Jarvis heard the dull thud against Jack's head and knew that he was falling. Whirling swiftly, he stopped a savage blow that was aimed at the stricken man, and with a back-handed cut laid the striker low.

"All right, Jack; keep down till the stars are gone." He stood with one sturdy leg on each side of Jack's body and his big club made a charmed circle about him. It was not more than twenty seconds before the wheels were out of Jack's head and he was on his feet again, though not quite steady.

Jack's fall had given courage to the gang, and they made a furious attack upon Jarvis, who was now alone and not a little impeded by the friend at his feet. As Jack struggled to his legs, a furious blow directed at him was parried by Jarvis's left arm,—his right being busy guarding his own head. The blow was a fearful one; it broke the small bone in the forearm, beat down the guard, and came with terrible force upon poor Jack's left shoulder, disabling it for a minute. At the same time Jarvis received a nasty blow across the face from an unexpected quarter. He was staggered by it, but he did not fall. Jack's right arm was good and very angry; a savage jab with his club into the face of the man who had struck Jarvis laid him low, and Jack grinned with satisfaction.

Things were going hard with the young men. They had, indeed, disqualified nine of the enemy; but there were still eight or ten more, and through hard work and harder knocks they had lost more than half their own fighting strength. At this rate they would be used up completely while there were still three or four of the enemy on foot. This was when they needed aid, and aid came.

No sooner had Lars found himself at liberty and with a club in his hands than he began to use it with telling effect. He attacked the outer circle, striking every head he could reach, and such was his sprightliness that four men fell headlong before the others became aware of this attack from the rear. This diversion came at the right moment, and proved effective. There were now but six of the enemy in fighting condition, and these six were more demoralized by the sudden and unknown element of a rear attack than by the loss of their thirteen comrades. They hesitated, and half turned to look, and two of them fell under the blows of Jack and Jarvis. As the rest turned to escape, the Swede's club felled one, and the other three ran for dear life. They did not escape, however, for the long legs of the young men were after them. Young blood is hot, and the savage fight that had been forced upon these boys had aroused all that was savage in them. In an instant they overtook two of the fleeing men, but neither could strike an enemy in the back. Throwing aside their clubs, each seized his enemy by the shoulder, turned him face to face and smote him sore, each after his fashion. Then they laughed, took hold of hands, and walked wearily back to the carriage. Jarvis's face was covered with blood, and Jack's neck and shoulders were drenched,—his wound had bled freely. Lars had relieved the ladies on the box after administering kicks and blows in generous measure to the dazed and crippled miners, who were crawling off the road or staggering along it. The Swede had a strain of fierce North blood which was not easily laid when once aroused, and he glared around the battle-field, hoping to find signs of resistance. When none were to be seen, he donned his coachman's coat and sat the box like a sphinx.

The girls went quickly forward to meet the men. They said little, but they put their hands on their battered champions in a way to make the heart of man glad. The men were flushed and proud, as men have been, and men will be, through all time, when they have striven savagely against other savages in the sight of their mistresses, and have gained the victory. Their bruises were numb with exultation and their wounds dumb with pride. There was no regret for blows given or received,—no sympathy for fallen foe. The male fights, in the presence of the female, with savage delight, from the lowest to the highest ranks of creation, and we must forgive our boys for some cruel exultation as they looked on the field of strife. Better feelings will come when the blood flows less rapidly in their veins!

"We must hurry home," said Jane, "and let papa mend you." Then she burst into tears. "Oh, I am so sorry and so frightened! Do you feel *very* bad, Jack? I know you are suffering dreadfully, Mr. Jarvis. Can't I do something for you?"

"My arm is bruised a bit," said Jarvis; "if you don't mind, you can steady it a little."

Jane's soft hands clasped themselves tenderly over Jarvis's great fist, and she felt relieved in the thought that she was doing something for her hero. She held the great right hand of Hercules tenderly, and Jarvis never let her know that it was the *left* arm that had been broken. She felt certain that he must be suffering agony, for ever and anon his fingers would close over hers with a spasmodic grip that sent a thrill of mixed joy and pain to her heart.

While I was bandaging the broken arm I saw the young lady going through some pantomimic exercises with her hands, as if seeking to revive the memory of some previous position; then her face blazed with a light, half pleasure and half shame, and she disappeared.

When the carriage arrived at Four Oaks, the story was told in few words, and I immediately set to work to "mend" the boys. Jack insisted that Jarvis should receive the first attention, and, indeed, he looked the worse. But after washing the blood off his face, I found that beyond a severe bruise, which would disfigure him for a few days, his face and head were unhurt. His arm was broken and badly contused. After I had attended to it, he said:—

"Doctor, I'm as good as new; hope Jack is no worse."

I carefully washed the blood off Jack's head and neck, and found an ugly scalp wound at least three inches long. It made me terribly anxious until I fairly proved that the bone was uninjured. After giving the boy the tonsure, I put six stitches into the scalp, and he never said a word. Perhaps the cause of this fortitude could be found in the blazing eyes of Jessie Gordon, which fixed his as a magnet, while her hands clasped his tightly. Miss Jessie was as white as snow, but there was no tremor in hand or eye. When it was all over, her voice was steady and low as she said:—

"Jack Williams, in the olden days men fought for women, and they were called knights. It was counted a noble thing to take peril in defence of the helpless. I find no record of more knightly deed than you have done to-day, and I know that no knight could have done it more nobly. I want you to wear this favor on your hand."

She kissed his hand and left the room. Jack didn't seem to mind the wound in his head, but he gave great attention to his hand.

CHAPTER XLIII

The Result

As soon as the first report of the battle reached me, I telephoned to Bill Jackson, asking him to come at once to Four Oaks and to bring a man with him. When he arrived, attended by his big Irishman, my men had already put one of the farm teams to a great farm wagon, and had filled the box nearly full of hay. We gave Jackson a hurried account of the fight and asked him to go at once and offer relief to the wounded,—if such relief were needed. Jackson was willing enough to go, but he was greatly disappointed that he had missed the fight; it seemed unnatural that there should be a big fight in his neighborhood and he not in it.

"I'd give a ten-acre lot to have been with you, lads," said the big farmer as he started off.

Word had been sent to Dr. High to be ready to care for some broken heads. Two hours later I drove to the Inn at Exeter and found the doctor just commencing the work of repair. Thirteen men had been brought in by the wagon, twelve of them more or less cut and bruised about the head, and all needing some surgical attention. The thirteenth man was stone dead. A terrific blow on the back of the head had crushed his skull as if it had been an egg-shell, and he must have died instantly. After looking this poor fellow over to make sure that there was no hope for him, we turned our attention to the wounded. The barn had been turned into a hospital, and in two hours we had a dozen sore heads well cared for, and their owners comfortably placed for the night on soft hay covered by blankets from the Inn. Mrs. French brought tea and gruels for the thirsty, feverish fellows, and we placed Otto and the big Irishman on duty as nurses for the night. The coroner had been summoned, and arrived as we finished our work. He was an energetic official, and lost no time in getting a jury of six to listen to the statements which the wounded men would give. To their credit be it said that every one who gave testimony at all, gave it to the effect that the miners were crazy-drunk, that they stopped the carriage, provoked the fight, and did their utmost to disable or destroy the enemy. The coroner would listen to no further testimony, but gave the case to the jury. In five minutes their verdict was returned, "justifiable and commendable homicide by person unknown to the jury."

The news of a fight and the death of a miner had reached Gordonville, where it created intense excitement. By the time the inquest was over a crowd of at least fifty miners had collected near the barn. Much grumbling and some loud threats were heard. Jackson took it upon himself to meet these angry men, and no one could have done better. Stepping upon a box which raised him a foot or two above the crowd, he said:—

"See here, fellows, I want to say a word to you. My name's Jackson—Bill Jackson; perhaps some of you know me. If you don't, I'll introduce myself. I wasn't in this fight,—worse luck for me! but I am wide open for engagements in that line. Some one inside said that this gang must be conciliated, and I thought I would come out and do it. I understand that you feel sore over this affair,—it's natural that you should,—but you must remember that those boys out at Four Oaks couldn't accommodate all of you. If you wouldn't mind taking me for a substitute, I'll do my level best to make it lively for you. You don't need cards of introduction to me; you needn't be American citizens; you needn't speak English; all you have to do is to put up your hands or cock your hats, and I'll know what you mean. If any of you thinks he hasn't had his share of what's been going on this afternoon, he may just call on Bill Jackson for the balance. I want to conciliate you if I can! I'm a good-tempered man, and not the kind to pick a quarrel; but if any of you low-lived dogs are looking for a fight, I'm not the man to disappoint you! I came out here to satisfy you in this matter and to send you home contented, and, by the jumping Jews! I'll do it if I have to break the head of every dog's son among you! They told me to speak gently to you, and by thunder, I've done it; but now I'm going to say a word for myself!

"A lot of your dirty crowd attacked two of the decentest men in the county when they were riding with ladies; one of the gang got killed and the rest got their skulls cracked. Would these boys fight for the girls they had with them? Hell's blazes! I'll fight for just thinking of it! Just one of you duffers say 'boo' to me! I'm going right through you!"

Jackson sprang into the crowd, which parted like water before a strong swimmer. He cocked his hat, smacked his fists, and invited any or all to stand up to him. He was crazy for a fight, to get even with Jack and Jarvis; but no one was willing to favor him. He marched through the gang lengthways, crossways, and diagonally, but to no purpose. In great disgust he returned to the barn and reported that the crowd would not be "conciliated." When we left, however, there were no miners to be seen.

It was after one o'clock in the morning when I reached home. Going directly to the room occupied by the boys, I met Polly on the stairs.

"I'm glad you've come," said she, "for I can't do a thing with those boys; they are too wild for any use."

Entering the room, I found the lads in bed, but hilarious. They had sent for Lars and had filled him full of hot stuff and commendation. He was sitting on the edge of a chair between the two beds, his honest eyes bulging and his head rolling from the effects of unusual potations. The lads had tasted the cup, too, but lightly; their high spirits came from other sources. Victories in war and in love deserve celebration; and when the two are united, a bit of freedom must be permitted. They sat bolt upright against the heads of their beds with flushed faces and shining eyes. They shouted Greek and Latin verse at the bewildered Swede; they gave him the story of Lars Porsena in the original, and then in bad Swedish. They called him Lars Porsena,—for had he not fought gallantly? Then he was Gustavus Adolphus,—for had he not come to the aid of the Protestants when they were in sore need? And then things got mixed and the "Royal Swede" was Lars Adolphus or Gustavus Porsena Viking all in one. The honest fellow was more than half crazed by strong waters, incomprehensible words, and "jollying up" which the young chaps had given him.

"See here, boys, don't you see that you're sending your noble Swede to his Lutzen before his time,—not dead, indeed, but dead drunk? This isn't the sort of medicine for either of you; you should have been asleep three hours ago. I'll take your last victim home."

We heard no more from any of the fighters until nine in the

morning. In looking them over I found that the Swede had as sore a head as either of the others, though he had never taken a blow.

Many friends came to see the boys during the days of their seclusion, to congratulate them on their fortunate escape, and to compliment them on their skill and courage. The lads enjoyed being made much of, and their convalescence was short and cheerful. Of course Sir Tom was the most constant and most enthusiastic visitor. The warm-hearted Irishman loved the boys always, but now he seemed to venerate them. The successful club fight appealed to his national instincts as nothing else could have done.

"With twenty years off and a shillalah in me hand I would have been proud to stand with you. By the Lord, I'm asking too much! I'll yield the twenty years and only ask for the stick!" And his cane went whirling around his head, now guarding, now striking, and now with elaborate flourishes, after the most approved Donny-brook fashion.

"But, me friend Jarvis, what is this you have on your face? Pond's Extract! Oh, murder! What is the world coming to when fresh beef and usquebaugh are crowded to the wall by bad-smelling water! Look at me nose; it is as straight as God made it, and yet many a time it has been knocked to one side of me face or spread all over me features. Nothing but whiskey and raw beef could ever coax it back! It's God's mercy if you are not deformed for life, me friend. Such privileges are not to be neglected with impunity. Let me bathe your face with whiskey and put a beef-steak poultice after it, and I'll have you as handsome as a girl in three days."

"Give me the steak and whiskey inside and I'll feel handsome at once," said Jarvis.

"Oh, the rashness of youth!" said Sir Tom. "But I'll not say a word against it. Youth is the greatest luck in the world, and I'll not copper it."

And then our sporting friend grew reminiscent and told of a time at Limmer's when the marquis and he occupied beds in the same room, not unlike our boys' room—only smoky and dingy—and poulticed their battered faces with beef, and used usquebaugh inside and outside, after ten friendly rounds.

"Queensbary's nose never resumed entirely after that night, but mine came back like rubber. Maybe it was the beef—maybe it was usquebaugh; me own preference is in favor of the latter."

Sir Tom came every day so long as the boys were confined to the place, and each day he was able to develop some new incident connected with the battle which called for applause. After hearing Lars tell his story for the fourth time, he gave him a ten-dollar note, saying:—

"You did nobly for a Swede, Mr. Gustavus Adolphus, but I would give ten tenners to have had your place and your shillalah,—a Swede for a match-lock, but an Irishman for a stick."

Jack had hardly recovered when he was waited on by a committee from the mine with a request that he would make another speech. He was asked to make good his offer of bonding the property, and also to formulate a plan of cooperation for the guidance of the men. Jack had the plans for a cooperative mining village well digested, and was anxious to get them before the miners. As soon as he was fit he went to Gordonville to try to organize the work. Jarvis of course went with him, and Bill Jackson and Sir Tom would not be denied; they did not say so, but they looked as if they thought some diversion might be found. In spite of the influence of strong whiskey, however, the meeting passed off peacefully. The results that grew from this effort at reformation were so great and so far-reaching that they deserve a book for their narration.

CHAPTER XLIV

Deep Waters

For sharp contrasts give me the dull country. The unexpected is the usual in small and in great things alike as they happen on a farm, and I make no apology to the reader for entering them in my narrative. I only ask him, if he be a city man, to take my word for the truth as to the general facts. To some elaboration and embellishment I plead guilty, but the groundwork is truth, and the facts stated are as real as the foundations of my buildings or the cows in my stalls. If the fortunate reader be a country man, he will need no assurance from me, for his eyes have seen and his ears have heard the strange and startling episodes with which the quiet country-side is filled. I do not dare record all the adventures which clustered around us at Four Oaks. People who know only the monotonous life of cities would not believe the half if told, and I do not wish to invite discredit upon my story of the making of the factory farm.

The incidents I have given of the strike at Gordon's mine are substantially correct, and I would love to follow them to their sequel,— the coöperative mine; but as that is a story by itself, I cannot do it now. I promise myself, however, the pleasure of writing a history of this innovation in coal-mining at an early date. It is worth the world's knowing that a copartnership can exist between three hundred equal partners without serious friction, and that community in business interests on a large scale can be successfully managed without any effort to control personal liberty, either domestic, social, or religious. Indeed, I believe the success of this experiment is due largely to the absence of any attempt to superintend the private interests of its members,—the only bond being a common financial one, and the one requisite to membership, ability to save a portion of the wages earned.

But to go back to farm matters. In August the ground was stirred for the second time around the young trees. To do this, the mulch was turned back and the surface for a space of three feet all around the tree was loosened by hoe or mattock, and the mulch was then returned. The trees were vigorous, and their leaves had the polish of health, in spite of the dry July and August. The mulching must receive the credit for much of this thrift, for it protected the soil from the rays of the sun and invited the deep moisture to rise toward the surface. Few people realize the amount of water that enters into the daily consumption of a tree. It is said that the four acres of leaf surface of a large elm will transpire or yield to evaporation eight tons of water in a day, and that it takes more than five hundred tons of water to produce one ton of hay, wheat, oats, or other crop. This seems enormous; but an inch of rain on an acre of ground means more than a hundred tons of water, and precipitation in our part of the country is about thirty-six inches per annum, so that we can count on over thirty-six hundred tons of water per acre to supply this tremendous evaporation of plant life.

Water-pot and hose look foolish in the face of these figures; indeed, they are poor makeshifts to keep life in plants during pinching times. A much more effective method is to keep the soil loose under a heavy mulch, for then the deep waters will rise. In our climate the tree's growth for the year is practically completed by July 15, and fortunately dry times rarely occur so early. We are, therefore, pretty certain to get the wood growth, no matter how dry the year, since it would take several years of unusual drought to prevent it. Of course the wood is not all that we wish for in fruit trees; the fruit is the main thing, and to secure the best development of it an abundant rainfall is needed after the wood is grown. If the rain doesn't come in July and August, heavy mulching must be the fruit-grower's reliance, and a good one it will prove if the drought doesn't continue more than one year. After July the new wood hardens and gets ready for the trying winter. If July and August are very wet, growth may continue until too late for the wood to harden, and it consequently goes into winter poorly prepared to resist its rigors. The result is a killing back of the soft wood, but usually no serious loss to the trees. The effort to stimulate late summer growth by cultivation and fertilization is all wrong; use manures and fertilizers freely from March until early June, but not later. The fall mulch of manure, if used, is more for warmth than for fertility; it is a blanket for the roots, but much of its value is leached away by the suns and rains of winter.

I felt that I had made a mistake in not sowing a cover crop in my orchard the previous year. There are many excellent reasons for the cover crop and not one against it. The first reason is that it protects the land from the rough usage and wash of winter storms; the second, that it adds humus to the soil; and the third, if one of the legumes is used, that it collects nitrogen from the air, stores it in each knuckle and joint, and holds it there until it is liberated by the decay of the plant. As nitrogen is the most precious of plant foods, and as the nitrate beds and deposits are rapidly becoming exhausted, we must look to the useful legumes to help us out until the scientists shall be able to fix the unlimited but volatile supply which the atmosphere contains, and thus to remove the certain, though remote, danger of a nitrogen famine. That this will be done in the near future by electric forces, and with such economy as to make the product available for agricultural purposes, is reasonably sure. In the meantime we must use the vetches, peas, beans, and clovers which are such willing workers.

The legumes fulfil the three requisites of the cover crop: protection, humus, and the storing of nitrogen. That was why, when the corn in the orchard was last cultivated in July, I planted cow peas between the rows. The peas made a fair growth in spite of the dry season, and after the corn was cut they furnished fine pasture for the brood sows, that ate the peas and trampled down the vines. In the spring ploughing this black mat was turned under, and with it went a store of fertility to fatten the land. Cow peas were sowed in all the corn land in 1897, and the rule of the farm is to sow corn-fields with peas, crimson clover, or some other leguminous plant. As my land is divided almost equally each year between corn and oats, which follow each other, it gets a cover crop turned under every two years over the whole of it. Great quantities of manure are hauled upon the oat stubble in the early spring, and these fields are planted to corn, while the corn stubble is fertilized by the cover crop, and oats are sown. The land is taxed heavily every year, but it increases in fertility and crop-making capacity. For the past two years my oats have averaged forty-seven bushels and my corn nearly sixty-eight bushels per acre. There is no waste land in my fields, and we have made such a strenuous fight against weeds that they no longer seriously tax the land. The wisdom of the work done on the fence rows is now apparent. The ploughing and seeding made it easy to keep the brush and weeds down; hay gathered close to the fences more than pays us for the mowing; and we have no tall weed heads to load the wind with seeds. This is a matter which is not sufficiently considered by the majority of farmers, for weeds are allowed to tax the land almost as much as crops do, and yet they pay no rent. Fence lines and corners are usually breeding beds for these pests, and it will pay any landowner to suppress them.

CHAPTER XLV

Dogs and Horses

It was definitely decided in August that Jane was not to go back to Farmington. We had all been of two minds over this question, and it was a comfort to have it settled, though I always suspect that my share of it was not beyond the suspicion of selfishness.

Jane was just past nineteen. She had a fair education, so far as books go, and she did not wish to graduate simply for the honor of a diploma. Indeed, there were many studies between her and the diploma which she loathed. She could never understand how a girl of healthy mind could care for mathematics, exact science, or dead languages. English and French were enough for her tongue, and history, literature, and metaphysics enough for her mind.

"I can learn much more from the books in your library and from the dogs and horses than I can at school, besides being a thousand times happier; and oh, Dad, if you will let me have a forge and work-shop, I will make no end of things."

This was a new idea to me, and I looked into it with some interest. I knew that Jane was deft with her fingers, but I did not know that she had a special wish to cultivate this deftness or to put it to practical use.

"What can you do with a forge?" said I. "You can't shoe the horses or sharpen the ploughs. Can you make nails? They are machine-made now, and you couldn't earn ten cents a week, even at horse-shoe nails."

"I don't want to make nails, Dad; I want to work in copper and brass, and iron, too, but in girl fashion. Mary Town has a forge in Hartford, and I spent lots of Saturdays with her. She says that I am cleverer than she is, but of course she was jollying me, for she makes beautiful things; but I can learn, and it's great fun."

"What kind of things does this young lady make, dear?"

"Lamp-shades, paper-knives, hinges, bag-tops, buckles, and lots of things. She could sell them, too, if she had to. It's like learning a trade, Dad."

"All right, child, you shall have a forge, if you will agree not to burn yourself up. Do you roll up your sleeves and wear a leather apron?"

"Why, of course, just like a blacksmith; only mine will be of soft brown leather and pinked at the edges."

So Jane was to have her forge. We selected a site for it at once in the grove to the east of the house and about 150 yards away, and set the carpenter at work. The shop proved to be a feature of the place, and soon became a favorite resort for old and young for five o'clock teas and small gossiping parties. The house was a shingled cottage, sixteen by thirty-two, divided into two rooms. The first room, sixteen by twenty, was the company room, but it contained a work bench as well as the dainty trappings of a girl's lounging room. In the centre of the wall that separated the rooms was a huge brick chimney, with a fireplace in the front room and a forge bed in the rear room, which was the forge proper.

I suppose I must charge the $460 which this outfit cost to the farm account and pay yearly interest on it, for it is a fixture; but I protest that it is not essential to the construction of a factory farm, and it may be omitted by those who have no daughter Jane.

There were other things hinging on Jane's home-staying which made me think that, from the standpoint of economy, I had made a mistake in not sending her back to Farmington. It was not long before the dog proposition was sprung upon me; insidiously at first, until I had half committed myself, and then with such force and sweep as to take me off my prudent feet. My own faithful terrier, which had dogged my heels for three years, seemed a member of the family, and reasonably satisfied my dog needs. That Jane should wish a terrier of some sort to tug at her skirts and claw her lace was no more than natural, and I was quite willing to buy a blue blood and think nothing of the $20 or $30 which it might cost. We canvassed the list of terriers,—bull, Boston, fox, Irish, Skye, Scotch, Airedale, and all,—and had much to say in favor of each. One day Jane said:—

"Dad, what do you think of the Russian wolf-hound?"

"Fine as silk," said I, not seeing the trap; "the handsomest dog that runs."

"I think so, too. I saw some beauties in the Seabright kennels.

Wouldn't one of them look fine on the lawn?—lemon and white, and so tall and silky. I saw one down there, and he wasn't a year old, but his tail looked like a great white ostrich feather, and it touched the ground. Wouldn't it be grand to have such a dog follow me when I rode. Say, Dad, why not have one?"

"What do you suppose a good one would cost?"

"I don't know, but a good bit more than a terrier, if they sell dogs by size. May I write and find out?"

"There's no harm in doing that," said I, like the jellyfish that I am.

Jane wasted no time, but wrote at once, and at least seventeen times each day, until the reply came, she gave me such vivid accounts of the beauties of the beasts and of the pleasure she would have in owning one, that I grew enthusiastic as well, and quite made up my mind that she should not be disappointed. When the letter came, there was suppressed excitement until she had read it, and then excitement unsuppressed.

"Dad, we can have Alexis, son of Katinka by Peter the Great, for $125! See what the letter says: 'Eleven months old, tall and strong in quarters, white, with even lemon markings, better head than Marksman, and a sure winner in the best of company.' Isn't that great? And I don't think $125 is much, do you?"

"Not for a horse or a house, dear, but for a dog—"

"But you know, Dad, this isn't a common dog. We mustn't think of it as a dog; it's a barzoi; that isn't too much for a barzoi, is it?"

"Not for a barzoi, or a yacht either; I guess you will have to have one or the other."

"The Seabright man says he has a girl dog by Marksman out of Katrina that is the very picture of Alexis, only not so large, and he will sell both to the same person for $200; they are such good friends."

"Break away, daughter, do you want a steam launch with your yacht?"

"But just think, Dad, only $75 for this one. You save $50, don't you see?"

"Dimly, I must confess, as through a glass darkly. But, dear, I may come to see it through your eyes and in the light of this altruistic dog fancier. I'm such a soft one that it's a wonder I'm ever trusted with money."

The natural thing occurred once more; the fool and his money parted company, and two of the most beautiful dogs came to live on our lawn. To live on our lawn, did I say? Not much! Such wonderful creatures must have a house and grounds of their own to retire to when they were weary of using ours, or when our presence bored them. The kennel and runs were built near the carriage barn, the runs, twenty by one hundred feet, enclosed with high wire netting. The kennel, eight by sixteen, was a handsome structure of its kind, with two compartments eight by eight (for Jane spoke for the future), and beds, benches, and the usual fixtures which well-bred dogs are supposed to require.

The house for these dogs cost $200, so I was obliged to add another $400 to the interest-bearing debt. "If Jane keeps on in this fashion," thought I, "I shall have to refund at a lower rate,"—and she did keep on. No sooner were the dogs safely kennelled than she began to think how fine it would look to be followed by this wonderful pair along the country roads and through the streets of Exeter. To be followed, she must have a horse and a saddle and a bridle and a habit; and later on I found that these things did not grow on the bushes in our neighborhood. I drew a line at these things, however, and decided that they should not swell the farm account. Thus I keep from the reader's eye some of the foolishness of a doting parent who has always been as warm wax in the hands of his, nearly always, reasonable children.

In my stable were two Kentucky-bred saddlers of much more than average quality, for they had strains of warm blood in their veins. There is no question nowadays as to the value of warm blood in either riding or driving horses. It gives ability, endurance, courage, and docility beyond expectation. One-sixteenth thorough blood will, in many animals, dominate the fifteen-sixteenths of cold blood, and prove its virtue by unusual endurance, stamina, and wearing capacity.

The blue-grass region of Kentucky has furnished some of the finest horses in the world, and I have owned several which gave grand service until they were eighteen or twenty years old. An honest horseman at Paris, Kentucky, has sold me a dozen or more, and I was willing to trust his judgment for a saddler for Jane. My request to him was for a light-built horse; weight, one thousand pounds; game and spirited, but safe for a woman, and one broken to jump. Everything else, including price, was left to him.

In good time Jane's horse came, and we were well pleased with it, as indeed we ought to have been. My Paris man wrote: "I send a bay mare that ought to fill the bill. She is as quiet as a kitten, can run like a deer, and jump like a kangaroo. My sister has ridden her for four months, and she is not speaking to me now. If you don't like her, send her back."

But I did like her, and I sent, instead, a considerable check. The mare was a bright bay with a white star on her forehead and white stockings on her hind feet, stood fifteen hands three inches, weighed 980 pounds, and looked almost too light built; but when we noted the deep chest, strong loins, thin legs, and marvellous thighs, we were free to admit that force and endurance were promised. Jane was delighted.

"Dad, if I live to be a hundred years old, I will never forget this day. She's the sweetest horse that ever lived. I must find a nice name for her, and to-morrow we will take our first ride, you and Tom and Aloha and I—yes, that's her name."

We did ride the next day, and many days thereafter; and Aloha proved all and more than the Kentuckian had promised.

CHAPTER XLVI

The Skim-Milk Trust

The third quarter of the year made a better showing than any previous one, due chiefly to the sale of hogs in August. The hens did well up to September, when they began to make new clothes for themselves and could not be bothered with egg-making. There were a few more than seven hundred in the laying pens, and nearly as many more rapidly approaching the useful age. The chief advantage in early chickens is that they will take their places at the nests in October or November while the older ones are dressmaking. This is important to one who looks for a steady income from his hens,—October and November being the hardest months to provide for. A few scattered eggs in the pullet runs showed that the late February and early March chickens were beginning to have a realizing sense of their obligations to the world and to the Headman, and that they were getting into line to accept them. More cotton-seed meal was added to the morning mash for the old hens, and the corn meal was reduced a little and the oatmeal increased, as was also the red pepper; but do what you will or feed what you like, the hen will insist upon a vacation at this season of the year. You may shorten it, perhaps, but you cannot prevent it. The only way to keep the egg-basket full is to have a lot of youngsters coming on who will take up the laying for October and November.

We milked thirty-seven cows during July, August, and September, and got more than a thousand pounds of milk a day. The butter sold amounted to a trifle more than $375 a month. I think this an excellent showing, considering the fact that the colony at Four Oaks never numbered less than twenty-four during that time, and often many more.

I ought to say that the calves had the first claim to the skim-milk; but as we never kept many for more than a few weeks, this claim was easily satisfied. It was like the bonds of a corporation,—the first claim, but a comparatively small one. The hens came next; they held preferred stock, and always received a five-pound, semi-daily dividend to each pen of forty. The growing pigs came last; they held the common stock, which was often watered by the swill and dish-water from both houses and the buttermilk and butter-washing from the dairy. I hold that the feeding value of skim-milk is not less than forty cents a hundred pounds, as we use it at Four Oaks. This seems a high price when it can often be bought for fifteen cents a hundred at the factories; but I claim that it is worth more than twice as much when fed in perfect freshness,—certainly $4 a day would not buy the skim-milk from my dairy, for it is worth more than that to me to feed. This by-product is essential to the smooth running of my factory. Without it the chickens and pigs would not grow as fast, and it is the best food for laying hens,—nothing else will give a better egg-yield. The longer my experiment continues, the stronger is my faith that the combination of cow, hog, and hen, with fruit as a filler, are ideal for the factory farm. With such a plant well-started and well-managed, and with favorable surroundings, I do not see how a man can prevent money from flowing to him in fair abundance. The record of the fourth quarter is as follows:—

Butter.	$1126.00
Eggs.	351.00
Hogs.	1807.00
Total	$3284.00

CHAPTER XLVII

Naboth's Vineyard

One hazy, lazy October afternoon, as my friend Kyrle and I sat on the broad porch hitting our pipes, sipping high balls, and watching the men and machines in the corn-fields, as all toiling sons of the soil should do, he said:—

"Doctor, I don't think you've made any mistake in this business."

"Lots of them, Kyrle; but none too serious to mend."

"Yes, I suppose so; but I didn't mean it that way. It was no mistake when you made the change."

"You're right, old man. It's done me a heap of good, and Polly and the youngsters were never so happy. I only wish we had done it earlier."

"Do you think I could manage a farm?"

"Why, of course you can; you've managed your business, haven't you? You've grown rich in a business which is a great sight more taxing. How have you done it?"

"By using my head, I suppose."

"That's just it; if a man will use his head, any business will go,—farming or making hats. It's the gray matter that counts, and the fellow that puts a little more of it into his business than his neighbor does, is the one who'll get on."

"But farming is different; so much seems to depend upon winds and rains and frosts and accidents of all sorts that are out of one's line."

"Not so much as you think, Kyrle. Of course these things cut in, but one must discount them in farming as in other lines of business. A total crop failure is an unknown thing in this region; we can count on sufficient rain for a moderate crop every year, and we know pretty well when to look for frosts. If a man will do well by his land, the harvest will come as sure as taxes. All the farmer has to do is to make the best of what Nature and intelligent cultivation will always produce. But he must use his gray matter in other ways than in just planning the rotation of crops. When he finds his raw staples selling for a good deal less than actual value,—less than he can produce them for, he should go into the market and buy against higher prices, for he may be absolutely certain that higher prices will come."

"But how is one to know? Corn changes so that one can't form much idea of its actual value."

"No more than other staples. You know what fur is worth, because you've watched the fur market for twenty years. If it should fall to half its present price, you would feel safe in buying a lot. You know that it would make just as good hats as it ever did, and that the hats, in all probability, would give you the usual profit. It's the same with corn and oats. I know their feeding value; and when they fall much below it, I fill my granary, because for my purpose they are as valuable as if they cost three times as much. Last year I bought ten thousand bushels of corn and oats at a tremendously low price. I don't expect to have such a chance again; but I shall watch the market, and if corn goes below thirty cents or oats below twenty cents, I will fill my granary to the roof. I can make them pay big profits on such prices."

"Will you sell this plant, Williams?"

"Not for a song, you may be sure."

"What has it cost you to date?"

"Don't know exactly,—between $80,000 and $90,000, I reckon; the books will show."

"Will you take twenty per cent advance on what the books show? I'm on the square."

"Now see here, old man, what would be the good of selling this factory for $100,000? How could I place the money so that it would bring me half the things which this farm brings me now? Could I live in a better house, or have better food, better service, better friends, or a better way of entertaining them? You know that $5000 or $6000 a year would not supply half the luxury which we secure at Four Oaks, or give half the enjoyment to my family or my friends. Don't you see that it makes little difference what we call our expenses out here, so long as the farm pays them and gives us a surplus besides? The investment is not large for one to get a living from, and it makes possible a lot of things which would be counted rank extravagance in the city.

Here's one of them."

A cavalcade was just entering the home lot. First came Jessie Gordon on her thoroughbred mare Lightfoot, and with her, Laura on my Jerry. Laura's foot is as dainty in the stirrup as on the rugs, and she has Jerry's consent and mine to put it where she likes. Following them were Jane and Bill Jackson, with Jane's slender mare looking absolutely delicate beside the big brown gelding that carried Jackson's 190 pounds with ease. The horses all looked as if there had been "something doing," and they were hurried to the stables. The ladies laughed and screamed for a season, as seems necessary for young ladies, and then departed, leaving us in peace. Jackson filled his pipe before remarking:—

"I've been over the ridge into the Dunkard settlement, and they have the cholera there to beat the band. Joe Siegel lost sixty hogs in three days, and there are not ten well hogs in two miles. What do you think of that?"

"That means a hard 'fight mit Siegel,'" said Kyrle.

"It ought to mean a closer quarantine on this side of the ridge," said I, "and you must fumigate your clothes before you appear before your swine, Jackson. It's more likely to be swine plague than cholera at this time of the year, but it's just as bad; one can hardly tell the difference, and we must look sharp."

"How does the contagion travel, Doctor?"

"On horseback, when such chumps as you can be found. You probably have some millions of germs up your sleeve now, or, more likely, on your back, and I wouldn't let you go into my hog pen for a $2000 note. I'm so well quarantined that I don't much fear contagion; but there's always danger from infected dust. The wind blows it about, and any mote may be an automobile for a whole colony of bacteria, which may decide to picnic in my piggery. This dry weather is bad for us, and if we get heavy winds from off the ridge, I'm going to whistle for rain."

"I say, Williams, when you came out here I thought you a tenderfoot, sure enough, who was likely to pay money for experience; but, by the jumping Jews! you've given us natives cards and spades."

"I *was* a tenderfoot so far as practical experience goes, but I tried to use the everyday sense which God gave me, and I find that's about all a man needs to run a business like this."

"You run it all right, for returns, and that's what we are after; and I'm beginning to catch on. I want you to tell me, before Kyrle here, why you gave me that bull two years ago."

"What's the matter with the bull, Jackson? Isn't he all right?"

"Sure he's all right, and as fine as silk; but why did you give him to me? Why didn't you keep him for yourself?"

"Well, Bill, I thought you would like him, and we were neighbors, and—"

"You thought I would save you the trouble of keeping him, didn't you?"

"Well, perhaps that did have some influence. You see, this is a factory farm from fence to fence, except this forty which Polly bosses, and the utilitarian idea is on top. Keeping the bull didn't exactly run with my notion of economy, especially when I could conveniently have him kept so near, and at the same time be generous to a neighbor."

"That's it, and it's taken me two years to find it out. You're trying to follow that idea all along the line. You're dead right, and I'm going to tag on, if you don't mind. I was glad enough for your present at the time, and I'm glad yet; but I've learned my lesson, and you may bet your dear life that no man will ever again give me a bull."

"That's right, Jackson. Now you have struck the key-note; stick to it, and you will make money twice as fast as you have done. Have a mark, and keep your eye on it, and your plough will turn a straight furrow."

Jackson sent for his horse, and just before he mounted, I said, "Are you thinking of selling your farm?"

"I used to think of it, but I've been to school lately and can 'do my sums' better. No, I guess I won't sell the paternal acres; but who wants to buy?"

"Kyrle, here, is looking for a farm about the size of yours, and to tell you the truth I should like him for a neighbor. It's dollars to doughnuts that I could give him a whole herd of bulls."

"Indeed, you can't do anything of the kind! I wouldn't take a gold dollar from you until I had it tested. I'm on to your curves."

"But seriously, Jackson, I must have more land; my stock will eat me out of house and home by the time the factory is running full steam. What would you say to a proposition of $10,000 for one hundred acres along my north line?"

"A year ago I would have jumped at it. Now I say 'nit.' I need it all, Doctor; I told you I was going to tag on. But what's the matter with the old lady's quarter across your south road?"

"Nothing's the matter with the land, only she won't sell it at any price."

"I know; but that drunken brute of a son will sell as soon as she's under the sod, and they say the poor old girl is on her last legs,—down with distemper or some other beastly disease. I'll tell you what I'll do. I'll sound the renegade son and see how he measures. Some one will get it before long, and it might as well be you."

Jackson galloped off, and Kyrle and I sat on the porch and divided the widow's 160-acre mite. It was a good strip of land, lying a fair mile on the south road and a quarter of a mile deep. The buildings were of no value, the fences were ragged to a degree, but I coveted the land. It was the vineyard of Naboth to me, and I planned its future with my friend and accessory sitting by. I destroyed the estimable old lady's house and barns, ran my ploughshares through her garden and flower beds, and turned the home site into one great field of lusty corn, without so much as saying by your leave. Thus does the greed of land grow upon one. But in truth, I saw that I must have more land. My factory would require more than ten thousand bushels of grain, with forage and green foods in proportion, to meet its full capacity, and I could not hope to get so much from the land then under cultivation. Again, in a few years—a very few—the fifty acres of orchard would be no longer available for crops, and this would still further reduce my tillable land. With the orchards out of use, I should have but 124 acres for all crops other than hay. If I could add this coveted 160, it would give me 250 acres of excellent land for intensive farming.

"I should like it on this side of the road," said I, "but I suppose that will have to do."

"What will have to do?" asked Kyrle.

"The 160 acres over there."

"You unconscionable wretch! Have you evicted the poor widow, and she on her deathbed? For stiffening the neck and hardening the heart, commend me to the close-to-nature life of the farmer. I wouldn't own a farm for worlds. It risks one's immortality. Give me the wicked city for pasturage—and a friend who will run a farm, at his own risk, and give me the benefit of it."

CHAPTER XLVIII

Maids and Mallards

We have so rarely entered our house with the reader that he knows little of its domestic machinery. So much depends upon this machinery that one must always take it into consideration when reckoning the pleasures and even the comforts of life anywhere, and this is especially true in the country. We have such a lot of people about that our servants cannot sing the song of lonesomeness that makes dolor for most suburbanites. They are "churched" as often as they wish, and we pay city wages; but still it is not all clear sailing in this quarter of Polly's realm. I fancy that we get on better than some of our neighbors; but we do not brag, and I usually feel that I am smoking my pipe in a powder magazine. There is something essentially wrong in the working-girl world, and I am glad that I was not born to set it right. We cannot down the spirit of unrest and improvidence that holds possession of cooks and waitresses, and we needs must suffer it with such patience as we can.

Two of our house servants were more or less permanent; that is, they had been with us since we opened the house, and were as content as restless spirits can be. These were the housekeeper and the cook,— the hub of the house. The former is a Norwegian, tall, angular, and capable, with a knot of yellow hair at the back of her head,—ostensibly for sticking lead pencils into,—and a disposition to keep things snug and clean. Her duties include the general supervision of both houses and the special charge of store-rooms, food cellars, and table supplies of all sorts. She is efficient, she whistles while she works, and I see but little of her. I suspect that Polly knows her well.

The cook, Mary, is small, Irish, gray, with the temper of a pepper-pod and the voice of a guinea-hen suffering from bronchitis, but she can cook like an angel. She is an artist, and I feel as if the seven-dollar-a-week stipend were but a "tip" to her, and that sometime she will present me with a bill for her services. My safeguard, and one that I cherish, is an angry word from her to the housekeeper. She jeeringly asserted that she, the cook, got $2 a week more than she, the housekeeper, did. As every one knows that the housekeeper has $5 a week, I am holding this evidence against the time when Mary asks for a lump sum adequate to her deserts. The number of things which Mary can make out of everything and out of nothing is wonderful; and I am fully persuaded that all the moneys paid to a really good cook are moneys put into the bank. I often make trips to the kitchen to tell Mary that "the dinner was great," or that "Mrs. Kyrle wants the receipt for that pudding," or that "my friend Kyrle asks if he may see you make a salad dressing;" but "don't do it, Mary; let the secret die with you." The cook cackles, like the guinea-hen that she is, but the dishes are none the worse for the commendation.

The laundress is just a washerwoman, so far as I know. She undoubtedly changes with the seasons, but I do not see her, though the clothes are always bleaching on the grass at the back of the house.

The maids are as changeable as old-fashioned silk. There are always two of them; but which two, is beyond me. I tell Polly that Four Oaks is a sprocket-wheel for maids, with two links of an endless chain always on top. It makes but little difference which links are up, so the work goes smoothly. Polly thinks the maids come to Four Oaks just as less independent folk go to the mountains or the shore, for a vacation, or to be able to say to the policeman, "I've been to the country." Their system is past finding out; but no matter what it is, we get our dishes washed and our beds made without serious inconvenience. The wage account in the house amounts to just $25 a week. My pet system of an increasing wage for protracted service doesn't appeal to these birds of passage, who alight long enough to fill their crops with our wild rice and celery, and then take wing for other feeding-grounds. This kind of life seems fitted for mallards and maids, and I have no quarrel with either. From my view, there are happier instincts than those which impel migration; but remembering that personal views are best applied to personal use, I wish both maids and mallards *bon voyage*.

CHAPTER XLIX

The Sunken Garden

Extending directly west from the porch for 150 feet is an open pergola, of simple construction, but fast gaining beauty from the rapid growth of climbers which Polly and Johnson have planted. It is floored with brick for the protection of dainty feet, and near the western end cluster rustic benches, chairs, tables, and such things as women and gardeners love. Facing the west 50 feet of this pergola is Polly's sunken flower garden, which is her special pride. It extends south 100 feet, and is built in the side of the hill so that its eastern wall just shows a coping above the close-cropped lawn. Of course the western wall is much higher, as the lawn slopes sharply; but it was filled in so as to make this wall-enclosed garden quite level. The walls which rise above the flower beds 4½ feet, are beginning to look decorated, thanks to creeping vines and other things which a cunning gardener and Polly know. Flowers of all sorts—annuals, biennials (triennials, perhaps), and perennials—cover the beds, which are laid out in strange, irregular fashion, far indeed from my rectangular style. These beds please the eye of the mistress, and of her friends, too, if they are candid in their remarks, which I doubt.

While excavating the garden we found a granite boulder shaped somewhat like an egg and nearly five feet long. It was a big thing, and not very shapely; but it came from the soil, and Polly wanted it for the base of her sun-dial. We placed it, big end down, in the mathematical centre of the garden (I insisted on that), and sunk it into the ground to make it solid; then a stone mason fashioned a flat space on the top to accommodate an old brass dial that Polly had found in Boston. The dial is not half bad. From the heavy, octagonal brass base rises a slender quill to cast its shadow on the figured circle, while around this circle old English characters ask, "Am I not wise, who note only bright hours?" A plat of sod surrounds the dial, and Polly goes to it at least once a day to set her watch by the shadow of the quill, though I have told her a hundred times that it is seventeen minutes off standard time. I am convinced that this estimable lady wilfully ignores conventional time and marks her cycles by such divisions as "catalogue time," "seed-buying time," "planting time," "sprouting time," "spraying time," "flowering time," "seed-gathering time," "mulching time," and "dreary time," until the catalogues come again. I know it seemed no time at all until she had let me in to the tune of $687 for the pergola, walls, and garden. She bought the sun-dial with her own money, I am thankful to say, and it doesn't enter into this account. I think it must have cost a pretty penny, for she had a hat "made over" that spring.

Polly has planted the lawn with a lot of shade trees and shrubs, and has added some clumps of fruit trees. Few trees have been planted near the house; the four fine oaks, from which we take our name, stand without rivals and give ample shade. The great black oak near the east end of the porch is a tower of strength and beauty, which is "seen and known of all men," while the three white oaks farther to the west form a clump which casts a grateful shade when the sun begins to decline. The seven acres of forest to the east is left severely alone, save where the carriage drive winds through it, and Polly watches so closely that the foot of the Philistine rarely crushes her wild flowers. Its sacredness recalls the schoolgirl's definition of a virgin forest: "One in which the hand of man has never dared to put his foot into it." Polly wanders in this grove for hours; but then she knows where and how things grow, and her footsteps are followed by flowers. If by chance she brushes one down, it rises at once, shakes off the dust, and says, "I ought to have known better than to wander so far from home."

She keeps a wise eye on the vegetable garden, too, and has stores of knowledge as to seed-time and harvest and the correct succession of garden crops. She and Johnson planned a greenhouse, which Nelson built, for flowers and green stuff through the winter, she said; but I think it is chiefly a place where she can play in the dirt when the weather is bad. Anyhow, that glass house cost the farm $442, and the interest and taxes are going on yet. I as well as Polly had to do some building that autumn. Three more chicken-houses were built, making five in all. Each consists in ten compartments twenty feet wide, of which each is intended to house forty hens. When these houses were completed, I had room for forty pens of forty each, which was my limit for laying hens. In addition was one house of ten pens for half-grown chickens and fattening fowls. It would take the hatch of another year to fill my pens, but one must provide for the future. These three houses cost, in round numbers, $2100,—five times as much as Polly's glass house,—but I was not going to play in them.

I also built a cow-house on the same plan as the first one, but about half the size. This was for the dry cows and the heifers. It cost $2230, and gave me stable room enough for the waiting stock, so that I could count on forty milch cows all the time, when my herd was once balanced. Forty cows giving milk, six hundred swine of all ages, putting on fat or doing whatever other duty came to hand, fifteen or sixteen hundred hens laying eggs when not otherwise engaged, three thousand apple trees striving with all their might to get large enough to bear fruit,—these made up my ideal of a factory farm; and it looked as if one year more would see it complete.

No rain fell in October, and my brook became such a little brook that I dared to correct its ways. We spent a week with teams, ploughs, and scrapers, cutting the fringe and frills away from it, and reducing it to severe simplicity. It is strange, but true, that this reversion to simplicity robbed it of its shy ways and rustic beauty, and left it boldly staring with open eyes and gaping with wide-stretched mouth at the men who turned from it. We put in about two thousand feet of tile drainage on both sides of what Polly called "that ditch," and this completed the improvements on the low lands. The land, indeed, was not too low to bear good crops, but it was lightened by under drainage and yielded more each after year.

The tiles cost me five cents per foot, or $100 for the whole. The work was done by my own men.

CHAPTER L

The Headman Generalizes

Jackson's prophecy came true. The old lady died, and before the ground was fairly settled around her the improvident son accepted a cash offer of $75 per acre for his homestead, and the farm was added to mine. This was in November. I at once spent $640 for 2-1/2 miles of fencing to enclose it in one field, charging the farm account with $12,640 for the land and fence.

This transaction was a bargain, from my point of view; and it was a good sale, from the standpoint of the other man, for he put $12,000 away at five per cent interest, and felt that he need never do a stroke of work again. A lazy man is easily satisfied.

In December I sold 283 hogs. It was a choice lot, as much alike as peas in a pod, and gave an average weight of 276 pounds; but the market was exceedingly low. I received the highest quotation for the month, $3.60 per hundred, and the lot netted $2702.

It seems hard luck to be obliged to sell fine swine at such a price, and a good many farmers would hold their stock in the hope of a rise; but I do not think this prudent. When a pig is 250 days old, if he has been pushed, he has reached his greatest profit-growth; and he should be sold, even though the market be low. If one could be certain that within a reasonable time, say thirty days, there would be a marked advance, it might do to hold; but no one can be sure of this, and it doesn't usually pay to wait. Market the product when at its best, is the rule at Four Oaks. The young hog is undoubtedly at his best from eight to nine months old. He has made a maximum growth on minimum feed, and from that time on he will eat more and give smaller proportionate returns. There is danger, too, that he will grow stale; for he has been subjected to a forcing system which contemplated a definite time limit and which cannot extend much beyond that limit without risks. Force your swine not longer than nine months and sell for what you can get, and you will make more money in the long run than by trying to catch a high market. I sold in December something more than four hundred cockerels, which brought $215. The apples from the old trees were good that year, but not so abundant as the year before, and they brought $337,—$2.25 per tree. The hens laid few eggs in October and November, though they resumed work in December; but the pullets did themselves proud. Sam said he gathered from fourteen to twenty eggs a day from each pen of forty, which is better than forty per cent. We sold nearly eighteen hundred dozen eggs during this quarter, for $553. The butter account showed nearly twenty-eight hundred pounds sold, which brought $894, and the sale of eleven calves brought $180. These sales closed the credit side of our ledger for the year.

Apples	$337.00
Calves	130.00
Cockerels	215.00
1785 doz. eggs	553.00
2790 lb. butter	894.00
283 hogs	2702.00
Total	$4831.00

In making up the expense account of that year and the previous one, I found that I should be able in future to say with a good deal of exactness what the gross amount would be, without much figuring. The interest account would steadily decrease, I hoped, while the wage account would increase as steadily until it approached $5500; that year it was $4662. Each man who had been on the farm more than six months received $18 more that year than he did the year before, and this increase would continue until the maximum wage of $40 a month was reached; but while some would stay long enough to earn the maximum, others would drop out, and new men would begin work at $20 a month. I felt safe, therefore, in fixing $5500 as the maximum wage limit of any year. Time has proven the correctness of this estimate, for $5372 is the most I have paid for wages during the seven years since this experiment was inaugurated.

The food purchased for cows, hogs, and hens may also be definitely estimated. It costs about $30 a year for each cow, $1 for each hog, and thirty cents for each hen. Everything else comes from the land, and is covered by such fixed charges as interest, wages, taxes, insurance, repairs, and replenishments. The food for the colony at

Four Oaks, usually bought at wholesale, doesn't cost more than $5 a month per capita. This seems small to a man who is in the habit of paying cash for everything that enters his doors; but it amply provides for comforts and even for luxuries, not only for the household, but also for the stranger within the gates. In the city, where water and ice cost money and the daily purchase of food is taxed by three or four middlemen, one cannot realize the factory farmer's independence of tradesmen. I do not mean that this sum will furnish terrapin and champagne, but I do not understand that terrapin and champagne are necessary to comfort, health, or happiness.

Let us look for a moment at some of the things which the factory farmer does not buy, and perhaps we shall see that a comfortable existence need not demand much more. His cows give him milk, cream, butter, and veal; his swine give roast pig, fresh pork, salt pork, ham, bacon, sausages, and lard; his hens give eggs and poultry; his fields yield hulled corn, samp, and corn meal; his orchards give apples, pears, peaches, quinces, plums, and cherries; his bushes give currants, gooseberries, strawberries, raspberries, blackberries; his vines give grapes; his forests give hickory nuts, butternuts, and hazel nuts; and, best of all, his garden gives more than twenty varieties of toothsome and wholesome vegetables in profusion. The whole fruit and vegetable product of the temperate zone is at his door, and he has but to put forth his hand and take it. The skilled housewife makes wonderful provision against winter from the opulence of summer, and her storehouse is crowded with innumerable glass cells rich in the spoils of orchard and garden. There is scant use for the grocer and the butcher under such conditions. I am so well convinced that my estimate of $5 a month is liberal that I have taxed the account with all the salt used on the farm.

CHAPTER LI

The Grand-Girls

The click of Jane's hammer began to be heard in November, and hardly a day passed without some music from this "Forge in the Forest." Sir Tom made a permanent station of the workshop, where he spent hours in a comfortable chair, drawing nourishment from the head of his cane and pleasure from watching the girl at the anvil. I suspect that he planted himself in the corner of the forge to safe-guard Jane; for he had an abiding fear that she would take fire, and he wished to be near at hand to put her out. He procured a small Bab-cock extinguisher and a half-dozen hand-grenades, and with these instruments he constituted himself a very efficient volunteer fire department. He made her promise, also, that she would have definite hours for heavy work, that he might be on watch; and so fond was she of his company, or rather of his presence, for he talked but little, that she kept close to the schedule.

Laura had a favorite corner in the forge, where she often turned a hem or a couplet. She was equally dexterous at either; and Sir Tom watched her, too, with an admiring eye. I once heard him say:—

"Milady Laura, it is the regret of me life that I came into the world a generation too soon."

Laura sometimes went away—she called it "going home," but we scoffed the term—and the doldrums blew until she returned. Sir Tom dined with us nearly every evening through the fall and early winter; and when he, and Kate and Tom and the grand-girls, and the Kyrles, and Laura were at Four Oaks, there was little to be desired. The grand-girls were nearly five and seven now, and they were a great help to the Headman. My terrier was no closer to my heels from morning to night than were these youngsters. They took to country life like the young animals they were, and made friends with all, from Thompson down. They must needs watch the sheep as they walked their endless way on the treadmill night and morning; they thrust their hands into hundreds of nests and placed the spoils in Sam's big baskets; they watched the calves at their patent feeders, which deceived the calves, but not the girls; they climbed into the grain bins and tobogganed on the corn; they haunted the cow-barn at milking time and wondered much; but the chiefest of their delights was the beautiful white pig which Anderson gave them. A little movable pen was provided for this favorite, and the youngsters fed it several times a day with warm milk from a nursing-bottle, like any other motherless child. The pig loved its foster-mothers, and squealed for them most of the time when it was not eating or sleeping; fortunately, a pig can do much of both. It grew playful and intelligent, and took on strange little human ways which made one wonder if Darwin were right in his conclusion that we are all ascended from the ape. I have seen features and traits of character so distinctly piggish as to rouse my suspicions that the genealogical line is not free from a cross of *sus scrofa*. The pig grew in stature and in wisdom, but not in grace, from day to day, until it threatened to dominate the place. However, it was lost during the absence of its friends,—to be replaced by a younger one at the next visit.

"Do *your* pigs get lost when you are away?" asked No. 1.

"Not often, dear."

"It's only pet pigs that runds away," said No. 2, "and I don't care, for it rooted me."

The pet pig is still a favorite with the grand-girls, but it always runs away in the fall.

Kate loved to come to Four Oaks, and she spent so much time there that she often said:—

"We have no right to that $1200; we spend four times as much time here as you all do in town."

"That's all right daughter, but I wish you would spend twice as much time here as you do, and I also wish that the $1200 were twice as much as it is."

Time was running so smoothly with us that we "knocked on wood" each morning for fear our luck would break.

The cottage which had once served as a temporary granary, and which had been moved to the building line two years before, was now turned into an overflow house against the time when Jack should come home for the winter vacation. Polly had decided to have "just as many as we can hold, and some more," and as the heaviest duties fell upon her, the rest of us could hardly find fault. The partitions were torn out of the cottage, and it was opened up into one room, except for the kitchen, which was turned into a bath-room. Six single iron beds were put up, and the place was made comfortable by an old-fashioned, air-tight, sheet-iron stove with a great hole in the top through which big chunks and knots of wood were fed. This stove would keep fire all night, and, while not up to latter-day demands, it was quite satisfactory to the warm-blooded boys who used it. The expense of overhauling the cottage was $214. Tom, Kate, and the grand-girls were to be with us, of course, and so were the Kyrles, Sir Tom, Jessie Gordon, Florence, Madeline, and Alice Chase. Jack was to bring Jarvis and two other men besides Frank and Phil of last year's party.

The six boys were bestowed in the cottage, where they made merry without seriously interrupting sleep in the main house. The others found comfortable quarters under our roof, except Sir Tom, who would go home some time in the night, to return before lunch the next day.

With such a houseful of people, the cook was worked to the bone; but she gloried in it, and cackled harder than ever. I believe she gave warning twice during those ten days; but Polly has a way with her which Mary cannot resist. I do not think we could have driven that cook out of the house with a club when there was such an opportunity for her to distinguish herself. Her warnings were simply matters of habit.

The holidays were filled with such things as a congenial country house-party can furnish—the wholesomest, jolliest things in the world; and the end, when it came, was regretted by all. I grew to feel a little bit jealous of Jarvis's attentions to Jane, for they looked serious, and she was not made unhappy by them. Jarvis was all that was honest and manly, but I could not think of giving up Jane, even to the best of fellows. I wanted her for my old age. I suspect that a loving father can dig deeper into the mud of selfishness than any other man, and yet feel all the time that he is doing God service. It is in accord with nature that a daughter should take the bit in her teeth and bolt away from this restraining selfishness, but the man who is left by the roadside cannot always see it in that light.

CHAPTER LII

The Third Reckoning

On the afternoon of December 31 I called a meeting of the committee of ways and means, and Polly and I locked ourselves in my office. It was then two and a half years since we commenced the experiment of building a factory farm, which was to supply us with comforts, luxuries, and pleasures of life, and yet be self-supporting: a continuous experiment in economics.

The building of the factory was practically completed, though not all of its machinery had yet been installed. We had spent our money freely,—too freely, perhaps; and we were now ready to watch the returns. Polly said:—

"There are some things we are sure of: we like the country, and it likes us. I have spent the happiest year of my life here. We've entertained more friends than ever before, and they've been better entertained, so that we are all right from the social standpoint. You are stronger and better than ever before, and so am I. Credit the farm with these things, Mr. Headman, and you'll find that it doesn't owe us such an awful amount after all."

"Are these things worth $100,000?"

"Now, John, you don't mean that you've spent $100,000! What in the world have you done with it? Just pigs and cows and chickens—"

"And greenhouses and sunken gardens and pergolas and kickshaws," said I. "But seriously, Polly, I think that we can show value for all that we have spent; and the whole amount is not three times what our city house cost, and that only covered our heads."

"How do you figure values here?"

"We get a great deal more than simply shelter out of this place, and we have tangible values, too. Here are some of them: 480 acres of excellent land, so well groomed and planted that it is worth of any man's money, $120 per acre, or $57,600; buildings, water-plant, etc., all as good as new, $40,000; 44 cows, $4400; 10 heifers nearly two years old, $500; 8 horses, $1200; 50 brood sows, $1000; 350 young pigs, $1700; 1300 laying hens, $1300; tools and machinery, $1500; that makes well over $100,000 in sight, besides all the things you mentioned before."

"You haven't counted the six horses in my barn."

"They haven't been charged to the farm, Polly."

"Or the trees you've planted?"

"No, they go with the land to increase its value."

"And my gardens, too?"

"Yes, they are fixtures and count with the acres. You see, this, land didn't cost quite $75 an acre, but I hold it $50 better for what we've done to it; I don't believe Bill Jackson would sell his for less. I offered him $10,000 for a hundred acres, and he refused. We've put up the price of real estate in this neighborhood, Mrs. Williams."

"Well, let's get at the figures. I'm dying to see how we stand."

"I have summarized them here:—

"To additional land and development of plant $20,353.00
To interest on previous investment 4,220.00
Wages 4,662.00
Food for twenty-five people 1,523.00
Food for stock. 2,120.00
Taxes and insurance 207.00
Shoeing and repairs 309.00

"Making in all $33,394.00

spent this year.

"The receipts are:—

"First quarter. $1,297.00
Second quarter 1,706.00
Third quarter 3,284.00
Fourth quarter. 4,831.00

"Making $11,118.00

"But we agreed to pay $4000 a year to the farm for our food and shelter, if it did as well by us as the town house did. Shall we do it, Polly?"

"Why, of course; we've been no end more comfortable here."

"Well, if we don't expect to get something for nothing, I think we ought to add it. Adding $4000 will make the returns from the farm $15,118, leaving $18,276 to add to the interest-bearing debt. Last year this debt was $84,404. Add this year's deficit, and we have $102,680. A good deal of money, Polly, but I showed you well over $100,000 in assets,—at our own price, to be sure, but not far wrong."

"Will you ever have to increase the debt?"

"I think not. I believe we shall reduce it a little next year, and each year thereafter. But, supposing it only pays expenses, how can you put on as much style on the interest of $100,000 anywhere else as you can here? It can't be done. When the fruit comes in and this factory is running full time, it will earn well on toward $25,000 a year, and it will not cost over $14,000 to run it, interest and all. It won't take long at that rate to wipe out the interest-bearing debt. You'll be rich, Polly, before you're ten years older."

"You are rich now, in imagination and expectation, Mr. Headman, but I'll bank with you for a while longer. But what's the use of charging the farm with interest when you credit it with our keeping?"

"There isn't much reason in that, Polly. It's about as broad as it is long. I simply like to keep books in that way. We charge the farm with a little more than $4000 interest, and we credit it with just $4000 for our food and shelter. We'll keep on in this way because I like it."

CHAPTER LIII

The Milk Machine

In opening the year 1898 I was faced by a larger business proposition than I had originally planned. When I undertook the experiment of a factory farm, I placed the limit of capital to be invested at about $60,000. Now I found that I had exceeded that amount by a good many thousand dollars, and I knew that the end was not yet. The factory was not complete, and it would be several years before it would be at its best in output. While it had cost me more than was originally contemplated, and while there was yet more money to be spent, there was still no reason for discouragement. Indeed, I felt so certain of ultimate profits that I was ready to put as much into it as could possibly be used to advantage.

The original plan was for a soiling farm on which I could milk thirty cows, fatten two hundred hogs, feed a thousand hens, and wait for thirty-five hundred fruit trees to come to a profitable age. With this in view, I set apart forty acres of high, dry land, for the feeding-grounds, twenty acres of which was devoted to the cows; and I now found that this twenty-acre lot would provide an ample exercise field for twice that number. It was in grass (timothy, red-top, and blue grass), and the cows nibbled persistently during the short hours each day when they were permitted to be on it; but it was never reckoned as part of their ration. The sod was kept in good condition and the field free from weeds, by the use of the mowing-machine, set high, every ten or twenty days, according to the season. Following the mower, we use a spring-tooth rake which bunched the weeds and gathered or broke up the droppings; and everything the rake caught was carted to the manure vats. Our big Holsteins do not suffer from close quarters, so far as I am able to judge, neither do they take on fat. From thirty minutes to three hours (depending on the weather), is all the outing they get each day; but this seems sufficient for their needs. The well-ventilated stable with its moderate temperature suits the sedentary nature of these milk machines, and I am satisfied with the results. I cannot, of course, speak with authority of the comparative merits of soiling *versus* grazing, for I have had no experience in the latter; but in theory soiling appeals to me, and in practice it satisfies me.

When I found I could keep more cows on the land set apart for them, I built another cow stable for the dry cows and the heifers, and added four stalls to my milk stable by turning each of the hospital wards into two stalls.

The ten heifers which I reserved in the spring of 1896 were now nearly two years old. They were expected to "come in" in the early autumn, when they would supplement the older herd. The cows purchased in 1895 were now five years old, and quite equal to the large demand which we made upon them. They had grown to be enormous creatures, from thirteen hundred to fourteen hundred pounds in weight, and they were proving their excellence as milk producers by yielding an average of forty pounds a day. We had, and still have, one remarkable milker, who thinks nothing of yielding seventy pounds when fresh, and who doesn't fall below twenty-five pounds when we are forced to dry her off. I have no doubt that she would be a successful candidate for advanced registration if we put her to the test. For ten months in each year these cows give such quantities of milk as would surprise a man not acquainted with this noble Dutch family. My five common cows were good of their kind, but they were not in the class with the Holsteins. They were not "robber" cows, for they fully earned their food; but there was no great profit in them. To be sure, they did not eat more than two-thirds as much as the Holsteins; but that fact did not stand to their credit, for the basic principle of factory farming is to consume as much raw material as possible and to turn out its equivalent in finished product. The common cows consumed only two-thirds as much raw material as the Holsteins, and turned out rather less than two-thirds of their product, while they occupied an equal amount of floor space; consequently they had to give place to more competent machines. They were to be sold during the season.

Why dairymen can be found who will pay $50 apiece for cows like those I had for sale (better, indeed, than the average), is beyond my method of reckoning values. Twice $50 will buy a young cow bred for milk, and she would prove both bread and milk to the purchaser in most cases. The question of food should settle itself for the dairyman as it does for the factory farmer. The more food consumed, the better for each, if the ratio of milk be the same.

My Holsteins are great feeders; more than 2 tons of grain, 2-1/2 tons of hay, and 4 or 5 tons of corn fodder, in addition to a ton of roots or succulent vegetables, pass through their great mouths each year. The hay is nearly equally divided between timothy, oat hay, and alfalfa; and when I began to figure the gross amount that would be required for my 50 Holstein gourmands, I saw that the widow's farm had been purchased none too quickly. To provide 100 tons of grain, 125 tons of hay, and 200 or 300 tons of corn fodder for the cows alone, was no slight matter; but I felt prepared to furnish this amount of raw material to be transmuted into golden butter. The Four Oaks butter had made a good reputation, and the four oak leaves stamped on each mould was a sufficient guarantee of excellence. My city grocer urged a larger product, and I felt safe in promising it; at the same time, I held him up for a slight advance in price. Heretofore it had netted me 32 cents a pound, but from January 1, 1898, I was to have 33-1/3 cents for each pound delivered at the station at Exeter, I agreeing to furnish at least 50 pounds a day, six days in a week.

This was not always easily done during the first eight months of that year, and I will confess to buying 640 pounds to eke out the supply for the colony; but after the young heifers came in, there was no trouble, and the purchased butter was more than made up to our local grocer.

It will be more satisfactory to deal with dairy matters in lump sums from now on. The contract with the city grocer still holds, and, though he often urges me to increase my herd, I still limit the supply to 300 pounds a week,—sometimes a little more, but rarely less. I believe that 38 to 44 cows in full flow of milk will make the best balance in my factory; and a well-balanced factory is what I am after.

I am told that animals are not machines, and that they cannot be run as such. My animals are; and I run them as I would a shop. There is no sentiment in my management. If a cow or a hog or a hen doesn't work in a satisfactory way, it ceases to occupy space in my shop, just as would an imperfect wheel. The utmost kindness is shown to all animals at Four Oaks. This rule is the most imperative one on the place, and the one in which no "extenuating circumstances" are taken into account. There are two equal reasons for this: the first is a deep-rooted aversion to cruelty in all forms; and the second is, *it pays*. But kindness to animals doesn't imply the necessity of keeping useless ones or those whose usefulness is below one's standard. If a man will use the intelligence and attention to detail in the management of stock that is necessary to the successful running of a complicated machine, he will find that his stock doesn't differ greatly from his machine. The trouble with most farmers is that they think the living machine can be neglected with impunity, because it will not immediately destroy itself or others, and because it is capable of a certain amount of self-maintenance; while the dead machine has no power of self-support, and must receive careful and punctual attention to prevent injury to itself and to other property. If a dairyman will feed his cows as a thresher feeds the cylinder of his threshing-machine, he will find that the milk will flow from the one about as steadily as the grain falls from the other.

Intensive factory farming means the use of the best machines pushed to the limit of their capacity through the period of their greatest usefulness, and then replaced by others. Pushing to the limit of capacity is in no sense cruelty. It is predicated on the perfect health of the animal, for without perfect condition, neither machine nor animal can do its best work. It is simply encouraging to a high degree the special function for which generations of careful breeding have fitted the animal.

That there is gratification in giving milk, no well-bred cow or mother will deny. It is a joyous function to eat large quantities of pleasant food and turn it into milk. Heredity impels the cow to do this, and it would take generations of wild life to wean her from it. As well say that the cataleptic trance of the pointer, when the game bird lies close and the delicate scent fills his nostrils, is not a joy to him, or that the Dalmatian at the heels of his horse, or the foxhound when Reynard's trail is warm, receive no pleasure from their specialties.

Do these animals feel no joy in the performance of service which is bred into their bones and which it is unnatural or freakish for them to lack? No one who has watched the "bred-for-milk" cow can doubt that the joys of her life are eating, drinking, sleeping, and giving milk. Pushing her to the limit of her capacity is only intensifying her life, though, possibly, it may shorten it by a year or two. While she lives she knows all the happiness of cow life, and knows it to the full. What more can she ask? She would starve on the buffalo grass which supports her half-wild sister, "northers" would freeze her, and the snow would bury her. She is a product of high cow-civilization, and as such she must have the intelligent care of man or she cannot do her best. With this care she is a marvellous machine for the making of the only article of food which in itself is competent to support life in man. If my Holsteins are not machines, they resemble them so closely that I will not quarrel with the name.

What is true of the cow, is true also of the pork-making machine that we call the hog. His wild and savage progenitor is lost, and we have in his place a sluggish animal that is a very model as a food producer. His three pleasures are eating, sleeping, and growing fat. He follows these pleasures with such persistence that 250 days are enough to perfect him. It can certainly be no hardship to a pig to encourage him in a life of sloth and gluttony which appeals to his taste and to my profit.

Custom and interest make his life ephemeral; I make it comfortable. From the day of his birth until we separate, I take watchful care of him. During infancy he is protected from cold and wet, and his mother is coddled by the most nourishing foods, that she may not fail in her duty to him. During childhood he is provided with a warm house, a clean bed, and a yard in which to disport himself, and is fed for growth and bone on skim-milk, oatmeal, and sweet alfalfa. During his youth, corn meal is liberally added to his diet, also other dainties which he enjoys and makes much of; and during his whole life he has access to clean water, and to the only medicine which a pig needs,—a mixture of ashes, charcoal, salt, and sulphur.

When he has spent 250 happy days with me, we part company with feelings of mutual respect,—he to finish his mission, I to provide for his successor.

My early plan was to turn off 200 of this finished product each year, but I soon found that I could do much better. One can raise a crop of hogs nearly as quickly as a crop of corn, and with much more profit, if the food be at hand. There was likely to be an abundance of food. I was more willing to sell it in pig skins than in any other packages. My plan was now to turn off, not 200 hogs each year, but 600 or more. I had 60 well-bred sows, young and old, and I could count on them to farrow at least three times in two years. The litters ought to average 7 each, say 22 pigs in two years; 60 times 22 are 1320, and half of 1320 is 660. Yes, at that rate, I could count on about 600 finished hogs to sell each year. But if my calculations were too high, I could easily keep 10 more brood sows, for I had sufficient room to keep them healthy.

The two five-acre lots, Nos. 3 and 5, had been given over to the brood sows when they were not caring for young litters in the brood-house. Comfortable shelters and a cemented basin twelve feet by twelve, and one foot deep, had been built in each lot. The water-pipe that ran through the chicken lot (No. 4) connected with these basins, as did also a drain-pipe to the drain in the north lane, so that it was easy to turn on fresh water and to draw off that which was soiled. Through this device my brood sows had access to a water bath eight inches deep, whenever they were in the fields. My hogs, young or old, have never been permitted to wallow in mud. We have no mud-holes at Four Oaks to grow stale and breed disease. The breeding hogs have exercise lots and baths, but the young growing and fattening stock have neither. They are kept in runs twenty feet by one hundred, in bunches of from twenty to forty, according to age, from the time they are weaned until they leave the place for good. This plan, which I did not intend to change, opened a question in my mind that gave me pause. It was this: Can I hope, even with the utmost care, to keep the house for growing and fattening swine free from disease if I keep it constantly full of swine?

The more I thought about it the less probable it appeared. The pig-house had cost me $4320. Another would cost as much, if not more, and I did not like to go to the expense unless it were necessary. I worked over this problem for several days, and finally came to the conclusion that I should never feel easy about my swine until I had two houses for them, besides the brood-house for the sows. I therefore gave the order to Nelson to build another swine-house as soon as spring opened. My plan was, and I carried it out, to move all the colonies every three months, and to have the vacant house thoroughly cleaned, sprayed with a powerful germicide, and whitewashed. The runs were to be turned over, when the weather would permit, and the ground sown to oats or rye.

The new house was finished in June, and the pigs were moved into it on July 1st with a lease of three months. My mind has been easy on the question of the health of my hogs ever since; and with reason, for there has been no epizoötic or other serious form of disease in my piggery, in spite of the fact that there are often more than 1200 pigs of all degrees crowded into this five-acre lot. The two pig-houses and the brood-house, with their runs, cover the whole of the lot, except the broad street of sixty feet just inside my high quarantine fence, which encloses the whole of it.

CHAPTER LIV

Bacon and Eggs

Each hog turned out from my piggery weighing 270 pounds or more, has eaten of my substance not less than 500 pounds of grain, 250 pounds of chopped alfalfa, 250 pounds of roots or vegetables, and such quantities of skimmed milk and swill as have fallen to his share. I could reckon the approximate cost of these foods, but I will not do so. All but the middlings and oil meal come from the farm and are paid for by certain fixed charges heretofore mentioned. The middlings and oil meal are charged in the "food for animals" account at the rate of $1 a year for each finished hog.

The truth is that a large part of the food which enters into the making of each 300 pounds of live pork, is of slow sale, and that for some of it there is no sale at all,—for instance, house swill, dish-water, butter-washings, garden weeds, lawn clippings, and all sorts of coarse vegetables. A hog makes half his growth out of refuse which has no value, or not sufficient to warrant the effort and expense of selling it. He has unequalled facilities for turning non-negotiable scrip into convertible bonds, and he is the greatest moneymaker on the farm. If the grain ration were all corn, and if there were a roadside market for it at 35 cents a bushel, it would cost $3.12; the alfalfa would be worth $1.45, and the vegetables probably 65 cents, under like conditions, making a total of $5.22 as a possible gross value of the food which the hog has eaten. The gross value of these things, however, is far above their net value when one considers time and expense of sale. The hog saves all this trouble by tucking under his skin slow-selling remnants of farm products and making of them finished assets which can be turned into cash at a day's notice.

To feed the hogs on the scale now planned, I had to provide for something like 7000 bushels of grain, chiefly corn and oats, 100 tons of alfalfa, and an equal amount of vegetables, chiefly sugar beets and mangel-wurzel. Certainly the widow's land would be needed.

The poultry had also outgrown my original plans, and I had built with reference to my larger views. There were five houses on the poultry lot, each 200 feet long, and each divided into ten equal pens. Four of these houses were for the laying hens, which were divided into flocks of 40 each; while the other house was for the growing chickens and for cockerels being fattened for market.

There were now on hand more than 1300 pullets and hens, and I instructed Sam to run his incubator overtime that season, so as to fill our houses by autumn. I should need 800 or 900 pullets to make our quota good, for most of the older hens would have to be disposed of in the autumn,—all but about 200, which would be kept until the following spring to breed from.

I believe that a three-year-old hen that has shown the egg habit is the best fowl to breed from, and it is the custom at Four Oaks to reserve specially good pens for this purpose. The egg habit is unquestionably as much a matter of heredity as the milk or the fat producing habit, and should be as carefully cultivated. With this end in view, Sam added young cockerels to four of his best-producing flocks on January 1, and by the 15th he was able to start his incubators.

Breeding and feeding for eggs is on the same principle as feeding and breeding for milk. It is no more natural for a hen to lay eggs for human consumption than it is for the robin to do so, or for the cow to give more milk than is sufficient for her calf. Man's necessity has made demands upon both cow and hen, and man's intelligence has converted individualists into socialists in both of these races. They no longer live for themselves alone. As the cow, under favorable conditions, finds pleasure in giving milk, so does the hen under like conditions take delight in giving eggs,—else why the joyous cackle when leaving her nest after doing her full duty? She gloats over it, and glories in it, and announces her satisfaction to the whole yard. It is something to be proud of, and the cackling hen knows it better than you or I. It can be no hardship to push this egg machine to the limit of its capacity. It adds new zest to the life of the hen, and multiplies her opportunities for well-earned self-congratulation.

Our hens are fed for eggs, and we get what we feed for. I said of my hens that I would not ask them to lay more than eight dozen eggs each year, and I will stick to what I said. But I do not reject voluntary contributions beyond this number. Indeed, I accept them with thanks, and give Biddy a word of commendation for her gratuity.

Eight dozen eggs a year will pay a good profit, but if each of my hens wishes to present me with two dozen more, I slip 62 cents into my pocket and say, "I am very much obliged to you, miss," or madam, as the case may be. Most of my hens do remember me in this substantial way, and the White Wyandottes are in great favor with the Headman.

The houses in which my hens live are almost as clean as the one I inhabit (and Polly is tidy to a degree); their food is as carefully prepared as mine, and more punctually served; their enemies are fended off, and they are never frightened by dogs or other animals, for the five-acre lot on which their houses and runs are built is enclosed by a substantial fence that prevents any interloping; book agents never disturb their siestas, nor do tree men make their lives hideous with lithographs of impossible fruit on improbable trees. Whether I am indebted to one or to all of these conditions for my full egg baskets, I am unable to say; but I do not purpose to make any change, for my egg baskets are as full as a reasonable man could wish. As nearly as I can estimate, my hens give thirty per cent egg returns as a yearly average—about 120 eggs for each hen in 365 days. This is more than I ask of them, but I do not refuse their generosity.

Every egg is worth, in my market, 2-1/2 cents, which means that the yearly product of each hen could be sold for $3. Something more than two thousand dozen are consumed by the home colony or the incubators; the rest find their way to the city in clean cartons of one dozen each, with a stencil of Four Oaks and a guarantee that they are not twenty-four hours old when they reach the middleman.

In return for this $3 a year, what do I give my hens besides a clean house and yard? A constant supply of fresh water, sharp grits, oyster shells, and a bath of road dust and sifted ashes, to which is added a pinch of insect powder. Twice each day five pounds of fresh skim-milk is given to each flock of forty. In the morning they have a warm mash composed of (for 1600 hens) 50 pounds of alfalfa hay cut fine and soaked all night in hot water, 50 pounds of corn meal, 50 pounds of oat meal, 50 pounds of bran, and 20 pounds of either meat meal or cotton-seed meal. At noon they get 100 pounds of mixed grains—wheat and buckwheat usually—with some green vegetables to pick at; and at night 125 to 150 pounds of whole corn. There are variations of this diet from time to time, but no radical change. I have read much of a balanced ration, but I fancy a hen will balance her own ration if you give her the chance.

Milk is one of the most important items on this bill of fare, and all hens love it. It should be fed entirely fresh, and the crocks or earthen dishes from which it is eaten should be thoroughly cleansed each day. Four ounces for each hen is a good daily ration, and we divide this into two feedings.

Our 1600 hens eat about 75 tons of grain a year. Add to this the 100 tons which 50 cows will require, 200 tons for the swine, and 25 tons for the horses, and we have 400 tons of grain to provide for the stock on the factory farm. Nearly a fourth of this, in the shape of bran, gluten meal, oil meal, and meat meal, must be purchased, for we have no way of producing it. For the other 300 tons we must look to the land or to a low market. Three hundred tons of mixed grains means something like 13,000 bushels, and I cannot hope to raise this amount from my land at present.

Fortunately the grain market was to my liking in January of 1898; and though there were still more than 7000 bushels in my granary, I purchased 5000 bushels of corn and as much oats against a higher market. The corn cost 27 cents a bushel and the oats 22, delivered at Exeter, the 10,000 bushels amounting to $2450, to be charged to the farm account.

I was now prepared to face the food problem, for I had more than 17,000 bushels of grain to supplement the amount the farm would produce, and to tide me along until cheap grain should come again, or until my land should produce enough for my needs. The supply in hand plus that which I could reasonably expect to raise, would certainly provide for three years to come, and this is farther than the average farmer looks into the future. But I claim to be more enterprising than an average farmer, and determined to keep my eyes open and to take advantage of any favorable opportunity to strengthen my position.

In the meantime it was necessary to force my trees, and to secure more help for the farm work. To push fruit trees to the limit of healthy growth is practical and wise. They can accomplish as much in growth and development in three years, when judiciously stimulated, as in five or six years of the "lick-and-a-promise" kind of care which they usually receive.

A tree must be fed first for growth and afterward for fruit, just as a pig is managed, if one wishes quick returns. To plant a tree and leave it to the tenderness of nature, with only occasional attention, is to make the heart sick, for it is certain to prove a case of hope deferred. In the fulness of time the tree and "happy-go-lucky" nature will prove themselves equal to the development of fruit; but they will be slow in doing it. It is quite as well for the tree, and greatly to the advantage of the horticulturist, to cut two or three years out of this unprofitable time. All that is necessary to accomplish this is: to keep the ground loose for a space around the tree somewhat larger than the spread of its branches; to apply fertilizers rich in nitrogen; to keep the whole of the cultivated space mulched with good barn-yard manure, increasing the thickness of the mulch with coarse stuff in the fall, so as to lengthen the season of root activity; and to draw the mulch aside about St. Patrick's Day, that the sun's rays may warm the earth as early as possible. Moderate pruning, nipping back of exuberant branches, and two sprayings of the foliage with Bordeaux mixture, to keep fungus enemies in check, comprise all the care required by the growing tree. This treatment will condense the ordinary growth of five years into three, and the tree will be all the better for the forcing.

As soon as fruit spurs and buds begin to show themselves, the treatment should be modified, but not remitted. Less nitrogen and more phosphoric acid and potash are to be used, and the mulch should *not* be removed in the early spring. The objects now are, to stimulate the fruit buds and to retard activity in the roots until the danger from late frosts is past. As a result of this kind of treatment, many varieties of apple trees will give moderate crops when the roots are seven, and the trunks are six years old. Fruit buds showed in abundance on many of my trees in the fall of 1897, especially on the Duchess and the Yellow Transparent, and I looked for a small apple harvest that year.

CHAPTER LV

The Old Time Farm-Hand

With all my industries thus increasing, the necessity for more help became imperative. French and Judson had their hands more than full in the dairy barns, and had to be helped out by Thompson. Anderson could not give the swine all the attention they needed, and was assisted by Otto, who proved an excellent swineherd. Sam had the aid of Lars's boys with the poultry, and very efficient aid it was, considering the time they could give to it. They had to be off with the market wagon at 7.40, and did not return from school until 4 P.M. Lars was busy in the carriage barn; and though we spared him as much as possible from driving, he had to be helped out by Johnson at such times as the latter could spare from his greenhouse and hotbeds. Zeb took care of the farm teams; but the winter's work of distributing forage and grain, getting up wood and ice, hauling manure, and so forth, had to be done in a desultory and irregular manner. The spring work would find us wofully behindhand if I did not look sharp. I had been looking sharp since January set in, and had experienced, for the first time, real difficulties in finding anything like good help. Hitherto I had been especially fortunate in this regard. I had met some reverses, but in the main good luck had followed me. I had nine good men who seemed contented and who were all saving money,—an excellent sign of stability and contentment. Even Lars had not fallen from grace but once, and that could hardly be charged against him, for Jack and Jarvis had tempted him beyond resistance; while Sam's nose was quite blanched, and he was to all appearances firmly seated on the water wagon. Really, I did not know what labor troubles meant until 1898, but since then I have not had clear sailing.

From my previous experience with working-men, I had formed the opinion that they were reasoning and reasonable human beings,—with peculiarities, of course; and that as a class they were ready to give good service for fair wages and decent treatment. In early life I had been a working-man myself, and I thought I could understand the feelings and sympathize with the trials of the laborer from the standpoint of personal experience. I was sorely mistaken. The laboring man of to-day is a different proposition from the man who did manual labor "before the war." That he is more intelligent, more provident, happier, or better in any way, I sincerely doubt; that he is restless, dissatisfied, and less efficient, I believe; that he is unreasonable in his demands and regardless of the interests of his employer, I know. There are many shining exceptions, and to these I look for the ultimate regeneration of labor; but the rule holds true.

I do not believe that the principles of life have changed in forty years. I do not believe that an intelligent, able-bodied man need be a servant all his life, or that industry and economy miss their rewards, or that there is any truth in the theory that men cannot rise out of the rut in which they happen to find themselves. The trouble is with the man, not with the rut. He spends his time in wallowing rather than in diligently searching for an outlet or in honestly working his way up to it. Heredity and environment are heavy weights, but industry and sobriety can carry off heavier ones. I have sympathy for weakness of body or mind, and patience for those over whom inheritance has cast a baleful spell; but I have neither patience nor sympathy for a strong man who rails at his condition and makes no determined effort to better it.

The time and money wasted in strikes, agitations, and arbitrations, if put to practical use, would better the working-man enough faster than these futile efforts do. I have no quarrel with unions or combinations of labor, so far as they have the true interests of labor for an object; but I do quarrel with the spirit of mob rule and the evidences of conspicuous waste, which have grown so rampant as to overshadow the helpful hand and to threaten, not the stability of society—for in the background I see six million conservative sons of the soil who will look to the stability of things when the time comes—but the unions themselves.

I remember my first summer on a farm. It lasted from the first day of April to the thirty-first day of October, and on the evening of that day I carried to my father $28, the full wage for seven months. I could not have spent one cent during that time, for I carried the whole sum home; but I do not remember that I was conscious of any want. The hours on the farm were not short; an eight-hour day would have been considered but a half-day. We worked from sun to sun, and I grew and knew no sorrow or oppression. The next year I received the munificent wage of $6 a month, and the following year, $8.

In after years, in brick-yards, sawmills, lumber woods, or harvest fields, there was no arbitrary limit put upon the amount of work to be done. If I chose to do the work of a man and a half, I got $1.50 for doing it, and it would have been a bold and sturdy delegate who tried to hold me from it. I felt no need of help from outside. I was fit to care for myself, and I minded not the long hours, the hard work, or the hard bed. This life was preliminary to a fuller one, and it served its use. I know what tired legs and back mean, and I know that one need not have them always if he will use the ordinary sense which God gives. Genius, or special cleverness, is not necessary to get a man out of the rut of hard manual labor. Just plain, everyday sense will do. But before I had secured the three men for whom I was in search, I began to feel that this common sense of which we speak so glibly is a rare commodity under the working-man's hat. I advertised, sent to agencies and intelligence offices, interviewed and inspected, consulted friends and enemies, and so generally harrowed my life that I was fit to give up the whole business and retire into a cave.

By actual count, I saw more than one hundred men, of all ages, sizes, and colors. Eight of these were tried, of whom five were found wanting. Early in February I had settled upon three sober men to add to our colony. As none of these lasted the year out, I may be forgiven for not introducing them to the reader. They served their purpose, and mine too, and then drifted on.

CHAPTER LVI

The Syndicate

I do not wish to take credit for things which gave me pleasure in the doing, or to appear altruistic in my dealings with the people employed at Four Oaks. I tell of our business and other relations because they are details of farm history and rightfully belong to these pages. If I dealt fairly by my men and established relations of mutual confidence and dependence, it was not in the hope that my ways might be approved and commended, but because it paid, in more ways than one. I wanted my men to have a lively interest in the things which were of importance to me, that their efforts might be intelligent and direct; and I was glad to enter into their schemes, either for pleasure or for profit, with such aid as I could give. Cordial understanding between employee and employer puts life into the contract, and disposes of perfunctory service, which simply recognizes a definite deed for a definite compensation. Uninterested labor leaves a load of hay in the field to be injured, just because the hour for quitting has come, while interested labor hurries the hay into the barn to make it safe, knowing that the extra half-hour will be made up to it in some other way.

It pays the farmer to take his help into a kind of partnership, not always in his farm, but always in his consideration. That is why my farm-house was filled with papers and magazines of interest to the men; that is why I spent many an evening with them talking over our industries; that is why I purchased an organ for them when I found that Mrs. French, the dairymaid, could play on it; that is why I talked economy to them and urged them to place some part of each month's wage in the Exeter Savings Bank; and that is why, early in 1898, I formulated a plan for investing their wages at a more profitable rate of interest. I asked each one to give me a statement of his or her savings up to date. They were quite willing to do this, and I found that the aggregate for the eight men and three women was $2530. Anderson, who saved most of his wages, had an account in a city savings bank, and did not join us in our syndicate, though he approved of it.

The money was made up of sums varying from $90, Lena's savings, to $460 owned by Judson, the buggy man. My proposition was this: Pool the funds, buy Chicago, Rock Island, and Pacific stock, and hold it for one or two years. The interest would be twice as much as they were getting from the bank, while the prospect of a decided advance was good. I said to them:—

"I have owned Chicago, Rock Island, and Pacific stock for more than three years. I commenced to buy at fifty-seven, and I am still buying, when I can get hold of a little money that doesn't have to go into this blessed farm. It is now eighty-one, and it will go higher. I am so sure of this that I will agree to take the stock from each or all of you at the price you pay for it at any time during the next two years. There is no risk in this proposition to you, and there may be a very handsome return."

They were pleased with the plan, and we formed a pool to buy thirty shares of stock. Thompson and I were trustees, and the certificate stood in our names; but each contributor received a pro-rata interest; Lena, one thirtieth; Judson, five-thirtieths; and the others between these extremes. The stock was bought at eighty-two. I may as well explain now how it came out, for I am not proud of my acumen at the finish. A little more than a year later the stock reached 122, and I advised the syndicate to sell. They were all pleased at the time with the handsome profit they had made, but I suspect they have often figured what they might have made "if the boss hadn't been such a chump," for we have seen the stock go above two hundred.

This was not the only enterprise in which our colony took a small share. The people at Four Oaks are now content to hold shares in one of the great trusts, which they bought several points below par, and which pay $1\frac{3}{4}$. per cent every three months. Even Lena, who held only one share of the C., R.I., & P. five years ago, has so increased her income-bearing property that she is now looked upon as a "catch" by her acquaintances. If I am correctly informed, she has an annual income of $105, independent of her wages.

CHAPTER LVII

The Death of Sir Tom

At 7.30 on the morning of March 16, Dr. High telephoned me that Sir Thomas O'Hara was seriously ill, and asked me to come at once. It took but a few minutes to have Jerry at the door, and, breasting a cold, thin rain at a sharp gallop, I was at my friend's door before the clock struck eight. Dr. High met me with a heavy face.

"Sir Tom is bad," said he, "with double pneumonia, and I am awfully afraid it will go hard with him."

I remembered that my friend's pale face had looked a shade paler than usual the evening before, and that there had been a pinched expression around the nose and mouth, as if from pain; but Sir Tom had many twinges from his old enemy, gout, which he did not care to discuss, and I took little note of his lack of fitness. He touched the brandy bottle a little oftener than usual, and left for home earlier; but his voice was as cheery as ever, and we thought only of gout. He was taken with a hard chill on his way home, which lasted for some time after he was put to bed; but he would not listen to the requests of William and the faithful cook that the doctor be summoned. At last he fell into a heavy sleep from which it was hard to rouse him, and the servants followed their own desire and called Dr. High. He came as promptly as possible, and did all that could be done for the sick man.

A hurried examination convinced me that Dr. High's opinion of the gravity of the case was correct, and we telephoned at once for a specialist from the city, and for a trained nurse. After a short consultation with Dr. High I reëntered my friend's room, and I fear that my face gave me away, for Sir Tom said:—

"Be a man, Williams, and tell the whole of it."

"My dear old man, this is a tough proposition, but you must buck up and make a game fight. We have sent for Dr. Jones and a nurse, and we will pull you through, sure."

"You will try, for sure, but I reckon the call has come for me to cash in me checks. When that little devil Frost hit me right and left in me chest last night, I could see me finish; and I heard the banshee in me sleep, and that means much to a Sligo man."

"Not to this Sligo man, I hope," said I, though I knew that we were in deep waters.

The wise man and the nurse came out on the 10.30 train, the nurse bringing comfort and aid, but the physician neither. After thoroughly examining the patient, he simply confirmed our fears.

"Serious disease to overcome, and only scant vital forces; no reasonable ground for hope."

Sir Tom gave me a smile as I entered the room after parting from the specialist.

"I've discounted the verdict," said he, "and the foreman needn't draw such a long face. I've had my fling, like a true Irishman, and I'm ready to pay the bill. I won't have to come back for anything, Williams; there's nothing due me; but I must look sharp for William and the old girl in the kitchen,—faithful souls,—for they will be strangers in a strange land. Will you send for a lawyer?"

The lawyer came, and a codicil to Sir Thomas's will made the servants comfortable for life. All that day and the following night we hung around the sick bed, hoping for the favorable change that never came. On the morning of the 17th it was evident that he would not live to see the sun go down. We had kept all friends away from the sick chamber; but now, at his request, Polly, Jane, and Laura were summoned, and they came, with blanched faces and tearful eyes, to kiss the brow and hold the hands of this dear man. He smiled with contentment on the group, and said:—

"Me friends have made such a heaven of this earth that perhaps I have had me full share."

"Sir Tom," said I, "shall I send for a priest?"

"A priest! What could I do with a priest? Me forebears were on the Orange side of Boyne Water, and we have never changed color."

"Would you like to see a clergyman?"

"No, no; just the grip of a friend's hand and these angels around me. Asking pardon is not me long suit, Williams, but perhaps the time has come for me to play it. If the good God will be kind to me I will thank Him, as a gentleman should, and I will take no advantage of His kindness; but if He cannot see His way clear to do that, I will take what is coming."

"Dear Sir Tom," said Jane, with streaming eyes, "God cannot be hard with you, who have been so good to every one."

"If there's little harm in me life, there's but scant good, too; I can't find much credit. Me good angel has had an easy time of it, more's the pity; but Janie, if you love me, Le Bon Dieu will not be hard on me. He cannot be severe with a poor Irishman who never stacked the cards, pulled a race, or turned his back on a friend, and who is loved by an angel."

I asked Sir Tom what we should do for him after he had passed away.

"It would be foine to sleep in the woods just back of Janie's forge, where I could hear the click of her hammer if the days get lonely; but there's a little castle, God save the mark, out from Sligo. Me forebears are there,—the lucky ones,—and me wish is to sleep with them; but I doubt it can be."

"Indeed it can be, and it shall be, too," said Polly. "We will all go with you, Sir Tom, when June comes, and you shall sleep in your own ground with your own kin."

"I don't deserve it, Mrs. Williams, indeed I don't, but I would lie easier there. That sod has known us for a thousand years, and it's the greenest, softest, kindest sod in all the world; but little I'll mind when the breath is gone. I'll not be asking that much of you."

"My dear old chap, we won't lose sight of you until that green sod covers the stanchest heart that ever beat. Polly is right. We'll go with you to Sligo,—all of us,—Polly and Jane and Jack and I, and Kate and the babies, too, if we can get them. You shall not be lonesome."

"Lonesome, is it? I'll be in the best of company. Me heart is at rest from this moment, and I'll wait patiently until I can show you Sligo. This is a fine country, Mrs. Williams, and it has given me the truest friends in all the world, but the ground is sweet in Sligo."

His breath came fainter and faster, and we could see that it would soon cease. After resting a few minutes, Sir Tom said:—

"Me lady Laura, do you mind that prayer song, the second verse?"

Laura's voice was sobbing and uncertain as it quavered:—

"Other refuge have I none,"

but it gained courage and persuasiveness until it filled the room and the heart of the man with,—

"Cover my defenceless head,
With the shadow of Thy wing."

A gentle smile and the relaxing of closed hands completed the story of our loss, though the real weight of it came days and months later.

It was long before we could take up our daily duties with anything like the familiar happiness. Something had gone out of our lives that could never be replaced, and only time could salve the wounds. The dear man who had gone was no friend to solemn faces, and living interests must bury dead memories; but it was a long time before the click of Jane's hammer was heard in her forge; not until Laura had said, "It will please *him*, Jane."

CHAPTER LVIII

Bacteria

January, February, and March passed with more than the usual snow and rain,—fully ten inches of precipitation; but the spring proved neither cold nor late. During these three months we sold butter to the amount of $1283, and $747 worth of eggs; in all, $2030.

The ploughs were started in the highest land on the 11th of April, and were kept going steadily until they had turned over nearly 280 acres.

I decided to put the whole of the widow's field into corn, lots 8, 12, and 15 (84 acres) into oats, and 50 acres of the orchards into roots and sweet fodder corn. Number 13 was to be sown with buckwheat as soon as the rye was cut for green forage. I decided to raise more alfalfa, for we could feed more to advantage, and it was fast gaining favor in my establishment. It is so productive and so nutritious that I wonder it is not more generally used by farmers who make a specialty of feeding stock. It contains as much protein as most grains, and is wholesome and highly palatable if properly cured. It should be cut just as it is coming into flower, and should be cured in the windrow. The leaves are the most nutritious part of the plant, and they are apt to fall off if the cutting be deferred, or if the curing be *done carelessly*.

Lot No. 9 was to be fitted for alfalfa as soon as the season would permit. First, it must receive a heavy dressing of manure, to be ploughed under. The ordinary plough was to be followed in this case by a subsoiler, to stir the earth as deep as possible. When the seed was sown, the land was to receive five hundred pounds an acre of high-grade fertilizer, and one hundred pounds an acre of infected soil.

The peculiar bacterium that thrives on congenial alfalfa soil is essential to the highest development of the plant. Without its presence the grass fails in its chief function—the storing of nitrogen—and makes but poor growth. When the alfalfa bacteria are abundant, the plant flourishes and gathers nitrogen in knobs and bunches in its roots and in the joints of its stems.

I sent to a very successful alfalfa grower in Ohio for a thousand pounds of soil from one of his fields, to vaccinate my field with. This is not always necessary,—indeed, it rarely is, for alfalfa seed usually carry enough bacteria to inoculate favorable soils; but I wished to see if this infected soil would improve mine. I have not been able to discover any marked advantage from its use; the reason being that my soil was so rich in humus and added manures that the colonies of bacteria on the seeds were quite sufficient to infect the whole mass. Under less favorable conditions, artificial inoculation is of great advantage.

Wonderful are the secrets of nature. The infinitely small things seem to work for us and the infinitely large ones appear suited to our use; and yet, perhaps, this is all "seeming" and "appearing." We may ourselves be simply more advanced bacteria, working blindly toward the solution of an infinite problem in which we are concerned only as means to an end.

"Why should the spirit of mortal be proud," until it has settled its relative position with both Sirius and the micro-organisms, or has estimated its stature by view-points from the bacterial world and from the constellation of Lyra. Until we have been able to compare opinions from these extremes, if indeed they be extremes, we cannot expect to make a correct estimate of our value in the economy of the universe. I fancy that we are apt to take ourselves too seriously, and that we will sometime marvel at the shadow which we did not cast.

CHAPTER LIX

Match-Making

The home lot took on a home look in the spring of 1898. The lawn lost its appearance of newness; the trees became acquainted with each other; the shrubs were on intimate terms with their neighbors, and broke into friendly rivalry of blossoms; the gardens had a settled-down look, as if they had come to stay; and even the wall flowers were enjoying themselves. These efforts of nature to make us feel at ease were thankfully received by Polly and me, and we voted that this was more like home than anything else we had ever had; and when the fruit trees put forth their promise of an autumn harvest in great masses of blossoms, we declared that we had made no mistake in transforming ourselves from city to country folk.

"Aristocracy is of the land," said Polly. "It always has been and always will be the source of dignity and stability. I feel twice as great a lady as I did in the tall house on B— Street."

"So you don't want to go back to that tall house, madam?"

"Indeed I don't. Why should I?"

"I don't know why you should, only I remember Lot's wife looked back toward the city."

"Don't mention that woman! She didn't know what she wanted. You won't catch me looking toward the city, except once a week for three or four hours, and then I hurry back to the farm to see what has happened in my garden while I've been away."

"But how about your friends, Polly?"

"You know as well as I that we haven't lost a friend by living out here, and that we've tied some of them closer. No, sir! No more city life for me. It may do for young people, who don't know better, but not for me. It's too restricted, and there's not enough excitement."

"Country life fits us like paper on the wall," said I, "but how about the youngsters? If we insist on keeping children, we must take them into our scheme of life."

"Of course we must, but children are an unknown quantity. They are x in the domestic problem, and we cannot tell what they stand for until the problem is worked out. I don't see why we can't find the value of x in the country as easily as in the city. They have had city and school life, now let them see country life; the x will stand for wide experience at least."

"Jane likes it thus far," said I, "and I think she will continue; but I don't feel so sure about Jack."

"You're as blind as a bat—or a man. Jane loves country life because she's young and growing; but there's a subconscious sense which tells her that she's simply fitting herself to be carried off by that handsome giant, Jim Jarvis. She doesn't know it, but it's the truth all the same, and it will come as sure as tide; and when it does come, her life will be run into other moulds than we have made, no matter how carefully."

"I wonder where this modern Hercules is most vulnerable. I'll slay him if I find him mousing around my Jane."

"You will slay nothing, Mr. Headman, and you know it; you will just take what's coming to you, as others have done since the world was young."

"Well, I give fair warning; it's 'hands off Jane,' for lo, these many years, or some one will be brewing 'harm tea' for himself."

"You bark so loud no one will believe you can bite," said this saucy, match-making mother.

"How about Jack?" said I. "Have you settled the moulds he is to be run in?"

"Not entirely; but I am not as one without hope. Jack will be through college in June, and will go abroad with us for July and August; he will be as busy as possible with the miners from the moment he comes back; he is much in love with Jessie, the Gordon's have no other child, the property is large, Homestead Farm is only three miles, and—"

"Slow up, Polly! Slow up! Your main line is all right, but your terminal facilities are bad. Jack is to be educated, travelled, employed, engaged, married, endowed with Homestead Farm, and all that; but you mustn't kill off the Gordons. I swing the red lantern in front of that train of thought. Let Jack and Jessie wait till we are through with Four Oaks and the Gordons have no further use for Homestead Farm, before thinking of coupling that property on to this."

"Don't be a greater goose than you can help," said Polly. "You know what I mean. Men are so short-sighted! Laura says, 'the Headman ought to have a small dog and a long stick'; but no matter, I'll keep an eye on the children, and you needn't worry about country life for them. They'll take to it kindly."

"Well, they ought to, if they have any appreciation of the fitness of things. Did you ever see weather made to order before? I feel as if I had been measured for it."

"It suits my garden down to the ground," said Polly, who hates slang.

"It was planned for the farmer, madam. If it happens to fit the rose-garden mistress, it is a detail for you to note and be thankful for, but the great things are outside the rose gardens. Look at that cornfield! A crow could hide in it anywhere."

"What have crows hiding got to do with corn, I'd like to know?"

"When I was a boy the farmers used to say, 'If it will cover a crow's back on the Fourth of July, it will make good corn,' and I am farmering with old saws when I can't find new ones."

"It's all of three weeks yet to the Fourth of July, and your corn will cover a turkey by that time."

"I hope so, but we shan't be here to see it, more's the pity, as Sir Tom would say."

"Do you know, Kate says she won't go over. She doesn't think it would pay for so short a trip. Why do you insist upon eight weeks?"

"Well, now, I like that! When did I ever insist on anything, Mrs. Williams? Not since I knew you well, did I? But be honest, Polly. Who has done the cutting down of this trip? You and the youngsters may stay as long as you please, but I will be back here September 1st unless the *Normania* breaks a shaft."

"I wish we could go *over* on a German boat. I hate the Cunarders."

"So do I, but we must land at Queenstown. We must put Sir Tom under the sod at that little castle out from Sligo. Then we can do Holland and Belgium, and have a week or ten days in London."

"That will be enough. I do hope Johnson will take good care of my flowers; it's the very most important time, you know, and if he neglects them—"

"He won't neglect them, Polly; even if he does, they can be easily replaced. But the hay harvest, now, that's different; if they spoil the timothy or cut the alfalfa too late!"

"Bother your alfalfa! What do I care for that? Kate's coming out with the babies, and I'm going to put her in full charge of the gardens. She'll look after them, I'm sure. I'll tell you another bit of news: Jim Jarvis is bound to go with us, Jack says, and he has asked if we'll let him."

"How long have you had that up your sleeve, young woman? I don't like it a little bit! That is why you talked so like an oracle a little while ago! What does Jane say?"

"She doesn't say much, but I think she wouldn't object."

"Of course she can't object. You sick a big brute of a man on to a little girl, and she don't dare object; but I'll feed him to the fishes if he worries her."

"To be sure you will, Mr. Ogre. Anybody would be sure of that to hear you talk."

"Don't chaff me, Polly. This is a serious business. If you sell my girl, I'm going to buy a new one. I'll ask Jessie Gordon to go with us and, if Jack is half the man I take him to be, he'll replenish our stock of girls before we get back."

"Who is match-making now?"

"I don't care what you call it. I shall take out letters of marque and reprisal. I won't raise girls to be carried off by the first privateer that makes sail for them, without making some one else suffer. If Jarvis goes, Jessie goes, that's flat."

"I think it will be an excellent plan, Mr. Bad Temper, and I've no doubt that we can manage it."

"Don't say 'we' when you talk of managing it. I tell you I'm entirely on the defensive until some one robs me, then I'll take what is my neighbor's if I can get it. If it were not for my promise to Sir Tom, I wouldn't leave the farm for a minute! And I would establish a quarantine against all giants for at least five years."

"You know you like Jarvis. He is one of the best."

"That's all right, Polly. He's as fine as silk, but he isn't fine enough for our Jane yet."

THE FAT OF THE LAND ∞ 231

CHAPTER LX

"I Told You So"

It may be the limitless horizon, it may be the comradery of confinement, it may be the old superstition of a plank between one and eternity, or it may be some occult influence of ship and ocean; but certain it is that there is no such place in all the world as a deck of a transatlantic liner for softening young hearts, until they lose all semblance of shape, and for melting them into each other so that out of twain there comes but one. I think Polly was pleased to watch this melting process, as it began to show itself in our young people, from the safe retreat of her steamer chair and behind the covers of her book. I couldn't find that she read two chapters from any book during the whole voyage, or that she was miserable or discontented. She just watched with a comfortable "I told you so" expression of countenance; and she never mentioned home lot or garden or roses, from dock to dock.

It is as natural for a woman to make matches as for a robin to build nests, and I suppose I had as much right to find fault with the one as with the other. I did not find fault with her, but neither could I understand her; so I fretted and fumed and smoked, and walked the deck and bet on everything in sight and out of sight, until the soothing influence of the sea took hold of me, and then I drifted like the rest of them.

No, I will not say "like the rest of them," for I could not forgive this waste of space given over to water. In other crossings I had not noted the conspicuous waste with any feeling of loss or regret; but other crossings had been made before I knew the value of land. I could not get away from the thought that it would add much to the wealth of the world if the mountains were removed and cast into the sea. Not only that, but it would curb to some extent the ragings of this same turbulent sea, which was rolling and tossing us about for no really good reason that I could discover. The Atlantic had lost much of its romance and mystery for me, and I wondered if I had ever felt the enthusiasm which I heard expressed on all sides.

"There she spouts!" came from a dozen voices, and the whole passenger list crowded the port rail, just to see a cow whale throwing up streams of water, not immensely larger than the streams of milk which my cow Holsteins throw down. The crowd seemed to take great pleasure in this sight, but to me it was profitless.

I have known the day when I could watch the graceful leaps and dives of a school of porpoises, as it kept with easy fin, alongside of our ocean greyhound, with pleasure unalloyed by any feeling of non-utility. But now these "hogs of the sea" reminded me of my Chester Whites, and the comparison was so much in favor of the hogs of the land, that I turned from these spectacular, useless things, to meditate upon the price of pork. Even Mother Carey's chickens gave me no pleasure, for they reminded me of a far better brood at home, and I cheerfully thanked the noble Wyandottes who were working every third day so that I could have a trip to Europe. To be sure, I had European trips before I had Wyandottes; to have them both the same year was the marvel.

Before we reached Queenstown, Jarvis had gained some ground by twice picking me out of the scuppers; but as I resented his steadiness of foot and strength of hand, it was not worth mentioning. I could see, however, that these feats were great in Jane's eyes. The double rescue of a beloved parent, from, not exactly a watery grave, but a damp scupper, would never be forgotten. The giant let her adore his manly strength and beauty, and I could only secretly hope that some wave—tidal if necessary—would take him off his feet and send him into the scuppers. But he had played football too long to be upset by a watery wave, and I was balked of my revenge.

Jack and Jessie were rather a pleasure to me than otherwise. They settled right down to the heart-softening business in such matter-of-fact fashion that their hearts must have lost contour before the voyage was half over. Polly dismissed them from her mind with a sigh of satisfaction, and I then hoped that she would find some time to devote to me, but I was disappointed. She assured me that those two were safely locked in the fold, but that she could not "set her mind at rest" until the other two were safe. After that she promised to take me in hand; whether for reward or for punishment left me guessing.

The six and a half days finally came to an end, and we debarked for Queenstown. The journey across Ireland was made as quickly as slow trains and a circuitous route would permit, and we reached Sligo on the second day. Sir Thomas's agent met us, and we drove at once to the "little castle out from Sligo." It proved to be a very old little castle, four miles out, overlooking the bay. It was low and flat, with thick walls of heavy stone pierced by a few small windows, and a broad door made of black Irish oak heavily studded with iron. From one corner rose a square tower, thirty feet or more in height, covered with wild vines that twined in and out through the narrow, unglazed windows.

Within was a broad, low hall, from which opened four rooms of nearly equal size. There was little evidence that the castle had been inhabited during recent years, though there was an ancient woman care-taker who opened the great door for us, and then took up the Irish peasant's wail for the last of the O'Haras. She never ceased her crooning except when she spoke to us, which was seldom; but she placed us at table in the state dining room, and served us with stewed kid, potatoes, and goat's milk. The walls of the dining room were covered with ancient pictures of the O'Haras, but none so recent as a hundred years. We could well believe Sir Tom's words, "the sod has known us for a thousand years," when we looked upon the score of pictures, each of which stood for at least one generation.

The agent told us that our friend had never lived at the castle, but that he had visited the place as a child, and again just before leaving for America. A wall-enclosed lot about two hundred feet square was "the kindest sod in all the world to an O'Hara," and here we placed our dear friend at rest with the "lucky ones" of his race. No one of the race ever deserved more "luck" than did our Sir Tom. The young clergyman who read the service assured us that he had found it; and our minds gave the same evidence, and our hearts said Amen, as we turned from his peaceful resting-place by the green waters of Sligo Bay.

Two days later we were comfortably lodged at The Hague, from which we intended to "do" the little kingdom of Holland by rail, by canal, or on foot, as we should elect.

CHAPTER LXI

The Belgian Farmer

Leaving Holland with regret, we crossed the Schelde into Belgium, the cockpit of Europe. It is here that one sees what intensive farming is like. No fences to occupy space, no animals roaming at large, nothing but small strips of land tilled to the utmost, chiefly by hand. Little machinery is used, and much of the work is done after primitive fashions; but the land is productive, and it is worked to the top of its bent.

The peasant-farmer soils his cows, his sheep, his swine, in a way that is economical of space and food, if not of labor, and manages to make a living and to pay rent for his twenty-acre strip of land. His methods do not appeal to the American farmer, who wastes more grain and forage each year than would keep the Netherlander, his family, and his stock; but there is a lesson to be learned from this sub-division and careful cultivation of land. Belgian methods prove that Mother Earth can care for a great many children if she be properly husbanded, and that the sooner we recognize her capacity the better for us.

Abandoned farms are not known in Belgium and France, though the soil has been cultivated for a thousand years, and was originally no better than our New England farms, and not nearly so good as hundreds of those which are practically given over to "old fields" in Virginia.

It is neglect that impoverishes land, not use. Intelligent use makes land better year by year. The only way to wear out land is to starve and to rob it at the same time. Food for man and beast may be taken from the soil for thousands of years without depleting it. All it asks in return is the refuse, carefully saved, properly applied, and thoroughly worked in to make it available. If, in addition to this, a cover crop of some leguminous plant be occasionally turned under, the soil may actually increase in fertility, though it be heavily cropped each year.

It would pay the young American farmer to study Belgian

methods, crude though they are, for the insight he could gain into the possibilities of continuous production. The greatest number of people to the square mile in the inhabited globe live in this little, ill-conditioned kingdom, and most of them get their living from the soil. It has been the battle-field of Europe: a thousand armies have harrowed it; human blood has drenched it from Liège to Ostend; it has been depopulated again and again. But it springs into new life after each catastrophe, simply because the soil is prolific of farmers, and they cannot be kept down. Like the poppies on the field of Waterloo, which renew the blood-red strife each year, the Belgian peasant-farmer springs new-born from the soil, which is the only mother he knows.

After two weeks in Holland, two in Belgium, and two in London, we were ready to turn our faces toward home.

We took the train to Southampton, and a small side-wheel steamer carried us outside Southampton waters, where we tossed about for thirty minutes before the *Normania* came to anchor. The wind was blowing half a gale from the north, and we were glad to get under the lee of the great vessel to board her.

The transfer was quickly made, and we were off for New York. The wind gained strength as the day grew old, but while we were in the Solent the bluff coast of Devon and Cornwall broke its force sufficiently to permit us to be comfortable on the port side of the ship.

As night came on, great clouds rolled up from the northwest and the wind increased. Darkness, as of Egypt, fell upon us before we passed the Lizard, and the only things that showed above the raging waters were the beacon lights, and these looked dim and far away. Occasionally a flash of lightning threw the waters into relief, and then made the darkness more impenetrable. As we steamed beyond the Lizard and the protecting Cornish coast, the full force of the gale, from out the Irish Sea, struck us. We were going nearly with it, and the good ship pitched and reared like an angry horse, but did not roll much. Pitching is harder to bear than rolling, and the decks were quickly vacated.

I turned into my stateroom soon after ten o'clock, and then happened a thing which will hold a place in my memory so long as I have one. I did not feel sleepy, but I was nervous, restless, and half sick. I lay on my lounge for perhaps half an hour, and then felt impelled to go on deck. I wrapped myself in a great waterproof ulster, pulled my storm cap over my ears, and climbed the companionway. Two or three electric bulbs in sheltered places on deck only served to make the darkness more intense. I crawled forward of the ladies' cabin, and, supporting myself against the donkey-engine, peered at the light above the crow's-nest and tried to think that I could see the man on watch in the nest. I did see him for an instant, when the next flash of lightning came, and also two officers on the bridge; and I knew that Captain Bahrens was in the chart house. When the next flash came, I saw the other lookout man making his short turns on the narrow space of bow deck, and was tempted to join him; why, I do not know. I crept past the donkey-engine, holding fast to it as I went, until I reached the iron gate that closes the narrow passage to the bow deck. With two silver dollars in my teeth I staggered across this rail-guarded plank, and when the next flash came I was sitting at the feet of the lookout man with the two silver dollars in my out-stretched hand. He took the money, and let me crawl forward between the anchors and the high bulwark of the bows.

The sensations which this position gave me were strange beyond description. Darkness was thick around me; at one moment I was carried upward until I felt that I should be lost in the black sky, and the next moment the downward motion was so terrible that the blacker water at the bottom of the sea seemed near. I cannot say that I enjoyed it, but I could not give it up.

When the great bow rose, I stood up, and, looking over the bulwark, tried to see either sky or water, but tried in vain, save when the lightning revealed them both. When the bow fell, I crouched under the bulwark and let the sea comb over me. How long I remained at this weird post, I do not know; but I was driven from it in such terror as I hope never to feel again.

An unusually large wave carried me nearer the sky than I liked to be, and just as the sharp bow of the great iron ship was balancing on its crest for the desperate plunge, a glare of lightning made sky and sea like a sheet of flame and curdled the blood in my veins. In the trough of the sea, under the very foot of the immense steamship, lay a delicate pleasure-boat, with its mast broken flush with its deck, and its helpless body the sport of the cruel waves.

The light did not last longer than it would take me to count five, but in that time I saw four figures that will always haunt me. Two sailors in yachting costume were struggling hopelessly with the tiller, and the wild terror of their faces as they saw the huge destruction that hung over them is simply unforgettable.

The other two were different. A strong, blond man, young, handsome, and brave I know, stood bareheaded in front of the cockpit. With a sudden, vehement motion he drew the head of a girl to his breast and held it there as if to shut out the horrible world. There was no fear in his face,—just pain and distress that he was unable to do more. I am thankful that I did not see the face of the girl. Her brown hair has floated in my dreams until I have cried out for help; what would her face have done?

In the twinkling of an eye it was over. I heard a sound as when one breaks an egg on the edge of a cup,—no more. I screamed with horror, ran across the guarded plank, climbed the gate, and fell headlong and screaming over the donkey-engine. Picking up my battered self, I shouted:

"Bahrens! Bahrens! for God's sake, help! Man overboard! Stop the ship!"

I reached the ladder to the bridge just as the captain came out of the chart house.

"For God's sake, stop the ship! You've run down a boat with four people! Stop her, can't you!"

"It can't be done, man. If we've run down a boat, it's all over with it and all in it. I can't risk a thousand lives without hope of saving one. This is a gale, Doctor, and we have our hands full."

I turned from him in horror and despair. I stumbled to my stateroom, dropped my wet clothing in the middle of the floor, and knew no more until the trumpet called for breakfast. The rush of green waters was pounding at my porthole; the experience of the night came back to me with horror; the reek of my wet clothes sickened my heart, and I rang for the steward.

"Take these things away, Gustav, and don't bring them back until they are dry and pressed."

"What things does the Herr Doctor speak for?"

"The wet things there on the floor."

"Excuse me, but I have seen no things wet."

"You Dutch chump!" said I, half rising, "what do you mean by saying—Well, I'll be damned!" There were my clothes, dry and folded, on the couch, and my ulster and cap on their hook, without evidence of moisture or use.

"Gustav, remind me to give you three rix-dollars at breakfast."

"Danke, Herr Doctor."

Of such stuff are dreams made. But I will know those terror-stricken sailors if I do not see them for a hundred years; and I am glad the dark-haired girl did not realize the horror, but simply knew that the man loved her; and I often think of the man who did the nice thing when no one was looking, and whose face was not terrorized by the crack of doom.

CHAPTER LXII

Home-Coming

Even Polly was satisfied with our young people before we entered New York Bay. If anything in their "left pulmonaries" had remained unsoftened during the voyage out and the comradery of the Netherlands, it was melted into non-resistance by the homeward trip. I could not long hold out against the evidence of happiness that surrounded me, and I gave a half-grudging consent that Jarvis and Jane might play together for the next three or four years, if they would not ask to play "for keeps" until those years had passed. They readily gave the promise, but every one knows how such promises are kept. The children wore me out in time, as all children do in all kinds of ways, and got their own ways in less than half the contract period. I cannot put my finger on any punishment that has befallen them for this lack of filial consideration, and I am fifteen-sixteenths reconciled.

I was downright glad that Jack "made good" with Jessie Gordon. She was the sort of girl to get out the best that was in him, and I was glad to have her begin early. Try as I might, I could not feel unhappy that beautiful September morning as we steamed up the finest waterway to the finest city in the world. Deny it who will, I claim that our Empire City and its environments make the most impressive human show. There is more life, vigor, utility, gorgeousness about it than can be found anywhere else; and it has the snap and elasticity of youth, which are so attractive. No man who claims the privilege of American citizenship can sail up New York Bay without feeling pride in his country and satisfaction in his birthright. One doesn't disparage other cities and other countries when he claims that his own is the best.

We were not specially badly treated at the custom-house,—no worse, indeed, than smugglers, thieves, or pirates would have been; and we escaped, after some hours of confinement, without loss of life or baggage, but with considerable loss of dignity. How can a self-respecting, middle-aged man (to be polite to myself) stand for hours in a crowded shed, or lean against a dirty post, or sit on the sharp edge of his open trunk, waiting for a Superior Being with a gilt band around his hat, without losing some modicum of dignity? And how, when this Superior Being calls his number and kicks his trunk, is he to know that he is a free-born American citizen and a lineal descendant of Roger Williams? The evidence is entirely from within. How is he to support a countenance and mien of dignity while the secrets of his chest are laid bare and the contents of his trunk dumped on the dirty floor? And how must his eyes droop and his face take on a hang-dog look when his second-best coat is searched for diamonds, and his favorite (though worn) pajamas punched for pearls.

There are concessions to be made for one's great and glorious country, and the custom-house is one of them. Perhaps we will do better sometime, and perhaps, though this is unlikely, the customs inspectors of the future will disguise themselves as gentlemen. We finally passed the inquisition, and, with stuffed trunks and ruffled spirits, took cabs for the station, and were presently within the protecting walls at Four Oaks, there to forget lost dignities in the cultivation of land and new ones.

CHAPTER LXIII

An Hundred Fold

Kate declared that she had had the time of her life during her nine weeks' stay at Four Oaks. "People here every day, and the house full over Sunday. We've kept the place humming," said she, "and you may be thankful if you find anything here but a mortgage. When Tom and I get rich, we are going to be farm people."

"Don't wait for that, daughter. Start your country home early and let it grow up with the children. It doesn't take much money to buy the land and to get fruit trees started. If Tom will give it his care for three hours a week, he will make it at least pay interest and taxes, and it will grow in value every year until you are ready to live on it. Think how our orchards would look now if we had started them ten years ago! They would be fit to support an average family."

"There, Dad, don't mount your hobby as soon as ever you get home. But we *have* had a good time out here. Do you really think farming is all beer and skittles?"

"It has been smooth sailing for me thus far, and I believe it is simply a business with the usual ups and downs; but I mean to make the ups the feature in this case."

"Are you really glad to get back to it? Didn't you want to stay longer?"

"I had a fine trip, and all that, but I give you this for true; I don't think it would make me feel badly if I were condemned to stay within forty miles of this place for the rest of my life."

"I can't go so far as that with you, Dad, but perhaps I may when I'm older."

"Yes, age makes a difference. At forty a man is a fool or a farmer, or both; at fifty the pull of the land is mighty; at sixty it has full possession of him; at seventy it draws him down with other forces than that which Newton discovered, and at eighty it opens for him and kindly tucks the sod around him. Mother Earth is no stepmother, but warm and generous to all, and I think a fellow is lucky who comes to her for long years of bounty before he is compelled to seek her final hospitality."

"But, Dad, we can't all be farmers."

"Of course not, and there's the pity of it; but almost every man can have a plot of ground on which each year he can grow some new thing, if only a radish or a leaf of lettuce, to add to the real wealth of the world. I tell you, young lady, that all wealth springs out of the ground. You think that riches are made in Wall Street, but they are not; they are only handled and manipulated. Stop the work of the farmer from April to October of any year, and Wall Street would be a howling wilderness. The Street makes it easier to exchange a dozen eggs for three spools of silk, or a pound of butter for a hat pin, but that's all; it never created half the intrinsic value of twelve eggs or sixteen ounces of butter. It's only the farmer who is a wealth producer, and it's high time that he should be recognized as such. He's the husbandman of all life; without him the world would be depopulated in three years. You don't half appreciate the profession which your Dad has taken up in his old age."

"That sounds all right, but I don't think the farmer would recognize himself from that description. He doesn't live up to his possibilities, does he?"

"Mighty few people do. A farmer may be what he chooses to be. He's under no greater limitations than a business or a professional man. If he be content to use his muscle blindly, he will probably fall under his own harrow. So, too, would the merchant or the lawyer who failed to use his intelligence in his business. The farmer who cultivates his mind as well as his land, uses his pencil as often as his plough, and mixes brains with brawn, will not fall under his own harrow or any other man's. He will never be the drudge of soil or of season, for to a large extent he can control the soil and discount the season. No other following gives such opportunity for independence and self-balance."

"Almost thou persuadest me to become a farmer," said Kate, as we left the porch, where I had been admiring my land while I lectured on the advantages of husbandry.

Polly came out of the rose garden, where she had been examining her flowers and setting her watch, and said:—

"Kate, you and the grand-girls must stay this month out, anyway.

It seems an age since we saw you last."

"All right, if Dad will agree not to fire farm fancies and figures at me every time he catches me in an easy-chair."

"I'll promise, but you don't know what you're missing."

Four Oaks looked great, and I was tempted to tramp over every acre of it, saying to each, "You are mine"; but first I had a little talk with Thompson.

"Everything has been greased for us this summer," said Thompson. "We got a bumper crop of hay, and the oats and corn are fine! I allow you've got fifty-five bushels of oats to the acre in those shocks, and the corn looks like it stood for more than seventy. We sold nine more calves the end of June, for $104. Mr. Tom must have a lot of money for you, for in August we sold the finest bunch of shoates you ever saw,—312 of them. They were not extra heavy, but they were fine as silk. Mr. Tom said they netted $4.15 per hundred, and they averaged a little over 260 pounds. I went down with them, and the buyers tumbled over each other to get them. I was mighty proud of the bunch, and brought back a check for $3407."

"Good for you, Thompson! That's the best sale yet."

"Some of the heifers will be coming in the last of this month or the first of next. Don't you want to get rid of those five scrub cows?"

"Better wait six weeks, and then you may sell them. Do you know where you can place them?"

"Jackson was looking at them a few days ago, and said he would give $35 apiece for them; but they are worth more."

"Not for us, Thompson, and not for him, either, if he saw things just right. They're good for scrubs; but they don't pay well enough for us, and if he wants them he can have them at that price about the middle of October."

The credit account for the second quarter of 1898 stood:—

23 calves	$270.00
Eggs	637.00
Butter	1314.00
Total	$2221.00

CHAPTER LXIV

Comfort Me with Apples

September added a new item to our list of articles sold; small, indeed, but the beginning of the fourth and last product of our factory farm,—fruit from our newly planted orchards. The three hundred plum trees in the chicken runs gave a moderate supply for the colony, and the dwarf-pear trees yielded a small crop; but these were hardly included in our scheme. I expected to be able, by and by, to sell $200 or $300 worth of plums; but the chief income from fruit would come from the fifty acres of young apple orchards.

I hope to live to see the time when these young orchards will bring me at least $5 a year for each tree; and if I round out my expectancy (as the life-insurance people figure it), I may see them do much better. In the interim the day of small things must not be despised. In our climate the Yellow Transparent and the Duchess do not ripen until early September, and I was therefore at home in time to gather and market the little crop from my six hundred trees. The apples were carefully picked, for they do not bear handling well, and the perfect ones were placed in half-bushel boxes and sent to my city grocer. Not one defective apple was packed, for I was determined that the Four Oaks stencil should be as favorably known for fruit as for other products.

The grocer allowed me fifty cents a box. "The market is glutted with apples, but not your kind," said he. "Can you send more?" I could not send more, for my young trees had done their best in producing ninety-six boxes of perfect fruit. Boxes and transportation came to ten cents for each box, and I received $38 for my first shipment of fruit.

I cannot remember any small sum of money that ever pleased me more,—except the $28 which I earned by seven months of labor in my fourteenth year; for it was "first fruits" of the last of our interlacing industries.

Thirty-eight dollars divided among my trees would give one cent to each; but four years later these orchards gave net returns of ninety cents for each tree, and in four years from now they will bring more than twice that amount. At twelve years of age they will bring an annual income of $3 each, and this income will steadily increase for ten or fifteen years. At the time of writing, February, 1903, they are good for $1 a year, which is five per cent of $20.

Would I take $20 apiece for these trees? Not much, though that would mean $70,000. I do not know where I could place $70,000 so that it would pay five per cent this year, six per cent next year, and twenty per cent eight or ten years from now. Of course, $70,000 would be an exorbitant price to pay for an orchard like mine; but it must be remembered that I am old and cannot wait for trees to grow.

If a man will buy land at $50 or $60 an acre, plant it to apple trees (not less than sixty-five to the acre), and bring these trees to an age when they will produce fruit to the value of $1.50 each, they will not have cost more than $1.50 per tree for the land, the trees, and the labor.

I am too old to begin over again, and I wish to see a handsome income from my experiment before my eyes are dim; but why on earth young men do not take to this kind of investment is more than I can see. It is as safe as government bonds, and infinitely safer than most mercantile ventures. It is a dignified employment, free from the ordinary risks of business; and it is not likely to be overdone. All one needs is energy, a little money, and a good bit of well-directed intelligence. This combination is common enough to double our rural population, relieve the congestion in trades and underpaid employments, and add immensely to the wealth of the country. If we can only get the people headed for the land, it will do much toward solving the vexing labor problems, and will draw the teeth of the communists and the anarchists; for no one is so willing to divide as he who cannot lose by division. To the man who has a plot of ground which he calls his own, division doesn't appeal with any but negative force. Neither should it, until all available lands are occupied. Then he must move up and make room for another man by his side.

The sales for the quarter ending September 30 were as follows:—

96 half-bushel boxes of apples. $38.00
9 calves 104.00

Eggs	543.00
Butter	1293.00
Hogs	3407.00
Total	$5385.00

This was the best total for any three months up to date, and it made me feel that I was getting pretty nearly out of the woods, so far as increasing my investment went.

Including my new hog-house and ten thousand bushels of purchased grain, the investment, thought I, must represent quite a little more than $100,000, and I hoped not to go much beyond that sum, for Polly looked serious when I talked of six figures, though she was reconciled to any amount which could be stated in five.

My buildings were all finished, and were good for many years; and if they burned, the insurance would practically replace them. My granary was full enough of oats and corn to provide for deficits of years to come; and my flocks and herds were now at their maximum, since Sam had turned more than eight hundred pullets into the laying pens. I began to feel that the factory would soon begin to run full time and to make material returns for its equipment. It would, of course, be several years before the fruit would make much showing, but I am a patient man, and could wait.

CHAPTER LXV

The End of the Third Year

"Polly," said I, on the evening of December 31, "let's settle the accounts for the year, and see how much we must credit to 'experience' to make the figures balance."

"Aren't you going to credit anything to health, and good times generally? If not, you don't play fair."

"We'll keep those things in reserve, to spring on the enemy at a critical moment; perhaps they won't be needed."

"I fancy you will have to bring all your reserves into action this time, Mr. Headman, for you promised to make a good showing at the end of the third year."

"Well, so I will; at least, according to my own estimate; but others may not see it as I do."

"Don't let others see it at all, then. The experiment is yours, isn't it?"

"Yes, for us; but it's more than a personal matter. I want to prove that a factory farm is sound in theory and safe in practice, and that it will fit the needs of a whole lot of farmers."

"I hardly think that 'a whole lot of farmers,' or of any other kind of people, will put $100,000 into a farm on any terms. Don't you think you've been a little extravagant?"

"Only on the home forty, Polly. I will expound this matter to you some time until you fall asleep, but not to-day. We have other business on hand. I want to give you this warning to begin with: you are not to jump to a conclusion or on to my figures until you have fairly considered two items which enter into this year's expense account. I've built an extra hog-house and have bought ten thousand bushels of grain, at a total expense of about $6000. Neither of these items was really needed this year; but as they are our insurance against disease and famine, I secured them early and at low prices. They won't appear in the expense account again,—at least, not for many years,—and they give me a sense of security that is mighty comforting."

"But what if Anderson sets fire to your piggery, or lightning strikes your granary,—how about the expense account then?"

"What do you suppose fire insurance policies are for? To paper the wall? No, madam, they are to pay for new buildings if the old ones burn up. I charge the farm over $200 a year for this security, and it's a binding contract."

"Well, I'll try and forget the $6000 if you'll get to the figures at once."

"All right. First, let me go over the statement for the last quarter of the year. The sales were: apples, from 150 old trees at $3 per tree, $450; 10 calves, $115; 360 hens and 500 cockerels, $430; 5 cows (the common ones, to Jackson) at $35 each, $175; eggs, $827; butter, $1311; and 281 hogs, rushed to market in December when only about eight months old and sold for $3.70 per hundred to help swell this account, $2649; making a total for the fourth quarter of $5957.

"The items of expense for the year were:—

"Interest on investment	$5,132.00
New hog-house	4,220.00
10,000 bu. of grain	2,450.00
Food for colony	5,322.00
Food for stock	1,640.00
Seeds and fertilizers	2,155.00
Insurance and taxes	730.00
Shoeing and repairs	349.00
Replenishments	450.00
"Total	$22,760.00

"The credit account reads: first quarter, $2030; second quarter, $2221; third quarter, $5387; fourth quarter, $5957; total, $15,595.

"If we take out the $6670 for the extra piggery and the grain, the expense account and the income will almost balance, even leaving out the $4000 which we agreed to pay for food and shelter. I think that's a fair showing for the three years, don't you?"

"Possibly it is; but what a lot of money you pay for wages. It's the largest item."

"Yes, and it always will be. I don't claim that a factory farm can be run like a grazing or a grain farm. One of its objects is to furnish well-paid employment to a lot of people. We've had nine men and two lads all the year, and three extra men for seven months, three women on the farm and five in the house,—twenty-two people to whom we've paid wages this year. Doesn't that count for anything? How many did we keep in the city?"

"Four,—three women and a man."

"Then we give employment to eighteen more people at equally good wages and in quite as wholesome surroundings. Do you realize, Polly, that the maids in the house get $1300 out of the $5300,—one quarter of the whole? Possibly there is a suspicion of extravagance on the home forty."

"Not a bit of it! You know that you proved to me that it cost us $5200 a year for board and shelter in the city, and you only credit the farm with $4000. That other $1200 would more than pay the extra wages. I really don't think it costs as much to live here as it did on B—Street, and any one can see the difference."

"You are right. If we call our plant an even $100,000, which at five per cent would mean $5000 a year,—where can you get house, lawns, woods, gardens, horses, dogs, servants, liberty, birds, and sun-dials on a wide and liberal scale for $5000 a year, except on a farm like this? You can't buy furs, diamonds, and yachts with such money anyhow or anywhere, so personal expenditures must be left out of all our calculations. No, the wage account will always be the large one, and I am glad it is so, for it is one finger of the helping hand."

"You haven't finished with the figures yet. You don't know what to add to our *permanent* investment."

"That's quickly done. *Nineteen thousand five hundred and ninety-five dollars* from twenty-two thousand seven hundred and sixty dollars leaves three thousand one hundred and sixty-five dollars to charge to our investment. I resent the word 'permanent,' which you underscored just now, for each year we're going to have a surplus to subtract from this interest-bearing debt."

"Precious little surplus you'll have for the next few years, with Jack and Jane getting married, and—"

"But, Polly, you can't charge weddings to the farm, any more than we can yachts and diamonds."

"I don't see why. A wedding is a very important part of one's life, and I think the farm ought to be *made* to pay for it."

"I quite agree with you; but we must add $3165 to the old farm debt, and take up our increased burden with such courage as we may. In round figures it is $106,000. Does that frighten you, Polly?"

"A little, perhaps; but I guess we can manage it. *You* would have been frightened three years ago if some one had told you that you would put $106,000 into a farm of less than five hundred acres."

"You're right. Spending money on a farm is like other forms of vice,—hated, then tolerated, then embraced. But seriously, a man would get a bargain if he secured this property to-day for what it has cost us. I wouldn't take a bonus of $50,000 and give it up."

"You'll hardly find a purchaser at that price, and I'm glad you can't, for I want to live here and nowhere else."

CHAPTER LXVI

Looking Backward

With the close of the third year ends the detailed history of the factory farm. All I wish to do further is to give a brief synopsis of the debit and credit accounts for each of the succeeding four years.

First I will say a word about the people who helped me to start the factory. Thompson and his wife are still with me, and they are well on toward the wage limit. Johnson has the gardens and Lars the stables, and Otto is chief swineherd. French and his wife act as though they were fixtures on the place, as indeed I hope they are. They have saved a lot of money, and they are the sort who are inclined to let well enough alone. Judson is still at Four Oaks, doing as good service as ever; but I fancy that he is minded to strike out for himself before long. He has been fortunate in money matters since he gave up the horse and buggy; he informed me six months ago that he was worth more than $5000.

"I shouldn't have had five thousand cents if I'd stuck to that darned old buggy," said he, "and I guess I'll have to thank you for throwing me down that day."

Zeb has married Lena, and a little cottage is to be built for them this winter, just east of the farm-house; and Lena's place is to be filled by her cousin, who has come from the old country.

Anderson and Sam both left in 1898,—poor, faithful Anderson because his heart gave out, and Sam because his beacon called him.

Lars's boys, now sixteen and eighteen, have full charge of the poultry plant, and are quite up to Sam in his best days. Of course I have had all kinds of troubles with all sorts of men; but we have such a strong force of "reliables" that the atmosphere is not suited to the idler or the hobo, and we are, therefore, never seriously annoyed. Of one thing I am certain: no man stays long at our farm-house without apprehending the uses of napkin and bath-tub, and these are strong missionary forces.

Through careful tilth and the systematic return of all waste to the land, the acres at Four Oaks have grown more fertile each year. The soil was good seven years ago, and we have added fifty per cent to its crop capacity. The amount of waste to return to the land on a farm like this is enormous, and if it be handled with care, there will be no occasion to spend much money for commercial fertilizers. I now buy fertilizers only for the mid-summer dressing on my timothy and alfalfa fields. The apple trees are very heavily mulched, even beyond the spread of their branches, with waste fresh from the vats, and once a year a light dressing of muriate of potash is applied. The trees have grown as fast as could be desired, and all of them are now in bearing. The apples from these young trees sold for enough last year to net ninety cents for each tree, which is more than the trees have ever cost me.

In 1898 these orchards yielded $38; in 1899, $165; in 1900, $530; in 1901, $1117. Seven years from the date of planting these trees, which were then three years old, I had received in money $4720, or $1200 more than I paid for the fifty acres of land on which they grew. If one would ask for better returns, all he has to do is to wait; for there is a sort of geometrical progression inherent in the income from all well-cared-for orchards, which continues in force for about fifteen years. There is, however, no rule of progress unless the orchards are well cared for, and I would not lead any one to the mistake of planting an orchard and then doing nothing but wait. Cultivate, feed, prune, spray, dig bores, fight mice, rabbits, aphides, and the thousand other enemies to trees and fruit, and do these things all the time and then keep on doing them, and you will win out. Omit all or any of them, and the chances are that you will fail of big returns.

But orcharding is not unique in this. Every form of business demands prompt, timely, and intelligent attention to make it yield its best. The orchards have been my chief care for seven years; the spraying, mulching, and cultivation have been done by the men, but I think I have spent one whole year, during the past seven, among my trees. Do I charge my orchards for this time? No; for I have gotten as much good from the trees as they have from me, and honors are easy. A meditative man in his sixth lustrum can be very happy with pruning-hook and shears among his young trees. If he cannot, I am sincerely sorry for him.

I have not increased my plant during the past four years. My stock consume a little more than I can raise; but there are certain things which a farm will not produce, and there are other things which one had best buy, thus letting others work their own specialties.

If I had more land, would I increase my stock? No, unless I had enough land to warrant another plant. My feeding-grounds are filled to their capacity from a sanitary point of view, and it would be foolish to take risks for moderate returns. If I had as much more land, I would establish another factory; but this would double my business cares without adding one item to my happiness. As it is, the farm gives me enough to keep me keenly interested, and not enough to tire or annoy me. So far as profits go, it is entirely satisfactory. It feeds and shelters my family and twenty others in the colony, and also the stranger within the gates, and it does this year after year without friction, like a well-oiled machine.

Not only this. Each year for the past four, it has given a substantial surplus to be subtracted from the original investment. If I live to be sixty-eight years of age, the farm will be my creditor for a considerable sum. I have bought no corn or oats since January, 1898. The seventeen thousand bushels which I then had in my granary have slowly grown less, though there has never been a day when we could not have measured up seven thousand or eight thousand bushels. I shall probably buy again when the market price pleases me, for I have a horror of running short; but I shall not sell a bushel, though prices jump to the sky.

I have seen the time when my corn and oats would have brought four times as much as I paid for them, but they were not for sale. They are the raw material, to be made up in my factory, and they are worth as much to me at twenty cents a bushel as at eighty cents. What would one think of the manager of a silk-thread factory who sold his raw silk, just because it had advanced in price? Silk thread would advance in proportion, and how does the manager know that he can replace his silk when needed, even at the advanced price?

When corn went to eighty cents a bushel, hogs sold for $8.25 a hundred, and my twenty-cent corn made pork just as fast as eighty-cent corn would have done, and a great deal cheaper.

Once I sold some timothy hay, but it was to "discount the season," just as I bought grain.

On July 18, 1901, a tremendous rain and wind storm beat down about forty acres of oats beyond recovery. The next day my mowing machines, working against the grain, commenced cutting it for hay. Before it was half cut, I sold to a livery-stable keeper in Exeter fifty tons of bright timothy for $600. The storm brought me no loss, for the horses did quite as well on the oat hay as they ever had done on timothy, and $600 more than paid for the loss of the grain.

During the first three years of my experiment hogs were very low,—lower, indeed, than at any other period for forty years. It was not until 1899 that prices began to improve. During that year my sales averaged $4.50 a hundred. In 1900 the average was $5.25, in 1901 it was $6.10, and in 1902 it was just $7. It will be readily appreciated that there is more profit in pork at seven cents a pound than at three and a half cents; but how much more is beyond me, for it cost no more to get my swine to market last year than it did in 1896. I charge each hog $1 for bran and shorts; this is all the ready money I pay out for him. If he weighs three hundred pounds (a few do), he is worth $10.50 at $3.50 a hundred, or $21 at $7 a hundred; and it is a great deal pleasanter to say $1 from $21, leaves $20, than to say $1 from $10.50 leaves $9.50.

Of course, $1 a head is but a small part of what the hog has cost when ready for market, but it is all I charge him with directly, for his other expenses are carried on the farm accounts. The marked increase in income during the past four years is wholly due to the advance in the price of pork and the increased product of the orchards. The expense account has not varied much.

The fruit crop is charged with extra labor, packages, and transportation, before it is entered, and the account shows only net returns. I have had to buy new machinery, but this has been rather evenly distributed, and doesn't show prominently in any year.

In 1900 I lost my forage barn. It was struck by lightning on June 13, and burned to the ground. Fortunately, there was no wind, and the rain came in such torrents as to keep the other buildings safe. I had to scour the country over for hay to last a month, and the expense of this, together with some addition to the insurance money, cost the farm $1000 before the new structure was completed. I give below the income and the outgo for the last four years:—

	INCOME	EXPENSES	TO THE GOOD
1899	$17,780.00	$15,420.00	$2,360.00
1900	19,460.00	16,480.00	2,980.00
1901	21,424.00	15,520.00	5,904.00
1902	23,365.00	15,673.00	7,692.00

Making a total to the good of $18,936.00

These figures cover only the money received and expended. They take no account of the $4000 per annum which we agreed to pay the farm for keeping us, so long as we made it pay interest to us. Four times $4000 are $16,000 which, added to $18,936, makes almost $35,000 to charge off from the $106,000 of original investment.

Polly was wrong when she spoke of it as a *permanent* investment. Four years more of seven-dollar pork and thrifty apple growth will make this balance of $71,000 look very small. The interest is growing rapidly less, and it will be but a short time before the whole amount will be taken off the expense account. When this is done, the yearly balance will be increased by the addition of $5000, and we may be able to make the farm pay for weddings, as Polly suggested.

CHAPTER LXVII

Looking Forward

I am not so opinionated as to think that mine is the only method of farming. On the contrary, I know that it is only one of several good methods; but that it is a good one, I insist. For a well-to-do, middle-aged man who was obliged to give up his profession, it offered change, recreation, employment, and profit. My ability to earn money by my profession ceased in 1895, and I must needs live at ease on my income, or adopt some congenial and remunerative employment, if such could be found. The vision of a factory farm had flitted through my brain so often that I was glad of the opportunity to test my theories by putting them into practice. Fortunately I had money, and to spare; for I had but a vague idea of what money would be needed to carry my experiment to the point of self-support. I set aside $60,000 as ample, but I spent nearly twice that amount without blinking. It is quite likely that I could have secured as good and as prompt returns with two-thirds of this expenditure. I plead guilty to thirty-three per cent lack of economy; the extenuating circumstances were, a wish to let the members of my family do much as they pleased and have good things and good people around them, and a somewhat luxurious temperament of my own.

Polly and I were too wise (not to say too old) to adopt farming as a means of grace through privations. We wanted the good there was in it, and nothing else; but as a secondary consideration I wished to prove that it can be made to pay well, even though one-third of the money expended goes for comforts and kickshaws.

It is not necessary to spend so much on a five-hundred-acre farm, and a factory farm need not contain so many acres. Any number of acres from forty to five hundred, and any number of dollars from $5000 to $100,000, will do, so long as one holds fast to the rules: good clean fences for security against trespass by beasts, or weeds; high tilth, and heavy cropping; no waste or fallow land; conscientious return to the land of refuse, and a cover crop turned under every second year; the best stock that money can buy; feed for product, not simply to keep the animals alive; force product in every way not detrimental to the product itself; maintain a strict quarantine around your animals, and then depend upon pure food, water, air, sunlight, and good shelter to keep them healthy; sell as soon as the product is finished, even though the market doesn't please you; sell only perfect product under your own brand; buy when the market pleases you and thus "discount the seasons"; remember that interdependent industries are the essence of factory farming; employ the best men you can find, and keep them interested in your affairs; have a definite object and make everything bend toward that object; plant apple trees galore and make them your chief care, as in time they will prove your chief dependence. These are some of the principles of factory farming, and one doesn't have to be old, or rich, to put them into practice.

I would exchange my age, money, and acres for youth and forty acres, and think that I had the best of the bargain; and I would start the factory by planting ten acres of orchard, buying two sows, two cows, and two setting hens. Youth, strength, and hustle are a great sight better than money, and the wise youth can have a finer farm than mine before he passes the half-century mark, even though he have but a bare forty to begin with.

I do not take it for granted that every man has even a bare forty; but millions of men who have it not, can have it by a little persistent self-denial; and when an able-bodied man has forty acres of ground under his feet, it is up to him whether he will be a comfortable, independent, self-respecting man or not.

A great deal of farm land is distant from markets and otherwise limited in its range of production, but nearly every forty which lies east of the hundredth meridian is competent to furnish a living for a family of workers, if the workers be intelligent as well as industrious. Farm lands are each year being brought closer to markets by steam and electric roads; telephone and telegraphic wires give immediate service; and the daily distribution of mails brings the producer into close touch with the consumer. The day of isolation and seclusion has passed, and the farmer is a personal factor in the market. He is learning the advantages of coöperation, both in producing and in disposing of his wares; he has paid off his mortgage and has money in the bank; he is a power in politics, and by far the most dependable element in the state. Like the wrestler of old, who gained new strength whenever his foot touched the ground, our country gains fresh vigor from every man who takes to the soil.

In preaching a hejira to the country, I do not forget the interests of the children. Let no one dread country life for the young until they come to the full pith and stature of maturity; for their chances of doing things worth doing in the world are four to one against those of children who are city-bred. Four-fifths of the men and women who do great things are country-bred. This is out of all proportion to the birth-rate as between country and city, and one is at a loss to account for the disproportion, unless it is to be credited to environment. Is it due to pure air and sunshine, making redder blood and more vigorous development, to broader horizons and freedom from abnormal conventions? Or does a close relation to primary things give a newness to mind and body which is granted only to those who apply in person?

Whatever the reason, it certainly pays to be country-bred. The cities draw to themselves the cream of these youngsters, which is only natural; but the cities do not breed them, except as exotics.

If the unborn would heed my advice, I would say, By all means be born in the country,—in Ohio if possible. But, if fortune does not prove as kind to you as I could wish, accept this other advice: Choose the, country for your foster-mother; go to her for consolation and rejuvenation, take her bounty gratefully, rest on her fair bosom, and be content with the fat of the land.

www.ingramcontent.com/pod-product-compliance
Lightning Source LLC
Chambersburg PA
CBHW031245090426
42742CB00007B/319